Liquid Crystal Optical Device

Liquid Crystal Optical Device

Special Issue Editors

Leszek R. Jaroszewicz
Noureddine Bennis

MDPI • Basel • Beijing • Wuhan • Barcelona • Belgrade

MDPI

Special Issue Editors
Leszek R. Jaroszewicz
Military University of Technology
Institute of Applied Physics
Poland

Noureddine Bennis
Military University of Technology
Institute of Applied Physics
Poland

Editorial Office
MDPI
St. Alban-Anlage 66
4052 Basel, Switzerland

This is a reprint of articles from the Special Issue published online in the open access journal *Crystals* (ISSN 2073-4352) in 2019 (available at: https://www.mdpi.com/journal/crystals/special_issues/Optical_Devices).

For citation purposes, cite each article independently as indicated on the article page online and as indicated below:

LastName, A.A.; LastName, B.B.; LastName, C.C. Article Title. *Journal Name* **Year**, *Article Number*, Page Range.

ISBN 978-3-03928-056-8 (Pbk)
ISBN 978-3-03928-057-5 (PDF)

Contents

About the Special Issue Editors

Leszek R. Jaroszewicz, PhD, DSc, Eng, SPIE Fellow is Director of the Institute of Applied Physics MUT. Since 1984, he has been engaged in the research of fiber optic coherent transmission, FOGs and interferometric and polarimetric optical fiber sensors, including the wide scope of the fiber optic Sagnac interferometer's applications as a sensor in a variety of physical fields. At present, his main field of interest is photonics technology application in sensors devices, including hybrid liquid crystal waveguide transducers, new technologies for monocrystals and glasses manufacturing especially of oxide type, the theory of complex semiconducting structures designed for their application in a new generation detectors, technologies of advanced fiber optics, as well as photonic crystal fiber elements. He is the author or co-author of more than 300 papers, 17 textbook contributions, and 12 patents as well as 20 patent applications.

Noureddine Bennis has worked in the field of liquid crystals since 2001. Prior to joining Military University of Technology (MUT) (Warsaw, Poland), he received his PhD in Physics from University of Valencia (Spain), and BS in Physics from University Abdelmalek Saadi of Tetuan (Morocco). He has authored more than 100 publications. Dr. Bennis has been working in liquid crystal (LC) photonic devices, with the overall objective of analyzing photonic devices that could be based on LC materials. His research at MUT focuses on adaptive lenses and the development of new class of liquid crystals for high-end photonic devices.

crystals

MDPI

Editorial
Liquid Crystal Optical Devices

Leszek R. Jaroszewicz * and Noureddine Bennis

Military University of Technology, Institute of Applied Physics, 00-908 Warsaw, Poland;
noureddine.bennis@wat.edu.pl
* Correspondence: leszek.jaroszewicz@wat.edu.pl

Received: 8 October 2019; Accepted: 9 October 2019; Published: 12 October 2019

It has been approximately 125 years since the Austrian scientist Friedrich Reinitzer in 1888 observed the curious behavior of the double melting points of cholesterol benzoate, a discovery that today is widely recognized as liquid crystal science. This discovery triggered a new area of research, engaging physicists and chemists around the world. The high optical anisotropy of liquid crystals implies large phase shifts in very short optical paths. Furthermore, their strong electro-optical effect allows for the rapid reorientation of their optical axis with, indeed, very low voltage in the range of only a few volts, hence, making liquid crystals compatible with current silicon technology [1]. Liquid crystal optical devices have provided the driving force for large amounts of research in photonics [2,3]. This technology has tremendous potential for technological breakthroughs in various fields and applications, from integrated optics [4] to detection and sensing [5]. The possibility to develop multifunctional macromolecular structures makes liquid crystals highly attractive candidates in the field of materials science and may represent an original strategy for the realization of molecular electronics-based devices [6].

As Guest Editors for the Special Issue "Liquid Crystal Optical Devices", we are pleased to present important contributions which are regularly submitted manuscripts, selected and reviewed via the regular system and accepted for publication. All papers presented here are based on original qualitative or quantitative research that opens new areas of inquiry and investigation in the field of liquid crystal optical devices.

The contents of this Special Issue reflect the rapid progress taking place in the field of liquid crystal devices. The first highlight of this thematic edition is an article entitled "Liquid Crystal Beam Steering Devices: Principles, Recent Advances, and Future Developments," by Ziqian He et al. [7], fellow researchers from the University of Central Florida, Orlando, USA. This article addresses the general operating principles of liquid crystal (LC) beam steering devices. The paper also focuses on two specific future challenges: fast response mid-infrared beam steering and device hybridization for large angle, high-efficiency beam steering.

The second highlight is an article entitled "Multifrequency Driven Nematics," by Noureddine Bennis et al. [8], fellow researchers from the Military University of Technology, Warsaw, Poland. This article addresses a novel LC mixture with frequency tunable capabilities. The tunability with frequency and the fast switching makes this LC of special interest for all kinds of optical phase modulators.

The third highlight is an article entitled "Recent Advances in Adaptive Liquid Crystal Lenses," by José Francisco Algorri et al. [9], fellow researchers from the University of Madrid, Leganés, Madrid, Spain. The authors reviewed recent advancements in adaptive LC lenses, introducing LC science and promising applications. Furthermore, novel applications of LC lenses were reviewed and the prospects and challenges of adaptive-focus LC lens technology were highlighted.

We anticipate that you will find all six articles presented in this special edition to be intriguing, thought provoking, and useful in reaching new milestones in your own research. Liquid Crystal Device is an important and interesting topic that we would like to keep attracting submissions in this field.

Crystals **2019**, *9*, 523

Now the editorial office of Crystals is running the second volume on this topic. Please recommend the journal *Crystals* to your colleagues and students to make this endeavor even more meaningful. All the papers published in this edition underwent a peer-reviewed process involving a minimum of two reviewers comprising internal as well as external referees.

We want to thank the authors for agreeing to publish their papers in this Special Issue, as well as the reviewers involved in the publishing process of these papers. We would also like to thank the *Crystals* publication Staff, who have produced a high-quality edition of this journal under the tight schedule required for this Special Issue.

We hope that this Special Issue will serve as a useful archival reference, providing access to information on liquid crystal optical devices.

References

1. Vettese, D. Liquid crystal on silicon. *Nat. Photonics* **2010**, *4*, 752–754. [CrossRef]
2. Smolyaninov, A.; El Amili, A.; Vallini, F.; Pappert, S.; Fainman, Y. Programmable plasmonic phase modulation of free-space wavefronts at gigahertz rates. *Nat. Photonics* **2019**, *13*, 431–435. [CrossRef]
3. Ford, A.D.; Morris, S.M.; Coles, H.J. Photonics and lasing in liquid-crystals. *Mater. Today* **2006**, *9*, 36–42. [CrossRef]
4. Tripathi, U.S.; Rastogi, V. Liquid crystal-based widely tunable integrated. *J. Opt. Soc. Am. B* **2019**, *36*, 1883–1889. [CrossRef]
5. Vallamkondu, J.; Corgiat, E.B.; Buchaiah, G.; Kandimalla, R.; Reddy, H. Liquid Crystals: A Novel Approach for Cancer Detection and Treatment. *Cancers* **2018**, *10*, 462. [CrossRef] [PubMed]
6. Gupta, R.K.; Sudhakar, A.A. Perylene-Based Liquid Crystals as Materials for Organic Electronics Applications. *Langmuir* **2019**, *35*, 2455–2479. [CrossRef] [PubMed]
7. Ziqian, H.; Fangwang, G.; Ran, C.; Kun, Y.; Zhan, Z.; Wu, S.T. Liquid Crystal Beam Steering Devices: Principles, Recent Advances, and Future Developments. *Crystals* **2019**, *9*, 292.
8. Bennis, N.; Herman, J.; Kalbarczyk, A.; Kula, P.; Jaroszewicz, L.R. Multifrequency Driven Nematics. *Crystals* **2019**, *9*, 275. [CrossRef]
9. Algorri, J.F.; Zografopoulos, D.C.; Urruchi, V.; Sánchez-Pena, J.M. Recent Advances in Adaptive Liquid Crystal Lenses. *Crystals* **2019**, *9*, 272. [CrossRef]

crystals

MDPI

Article

Monochromatic Depolarizer Based on Liquid Crystal

Paweł Marć *, Noureddine Bennis, Anna Spadło, Aleksandra Kalbarczyk, Rafał Węgłowski, Katarzyna Garbat and Leszek R. Jaroszewicz

Faculty of New Technologies and Chemistry, Military University of Technology, 2 gen. S. Kaliskiego St., 00-908 Warsaw, Poland
* Correspondence: pawel.marc@wat.edu.pl; Tel.: +48-261-839-424

Received: 24 May 2019; Accepted: 22 July 2019; Published: 28 July 2019

Abstract: Polarization is a very useful parameter of a light beam in many optical measurements. Improvement of holographic systems requires optical elements which need a diffused and depolarized light beam. This paper describes a simple monochromatic depolarizer based on a pure vertically aligned liquid crystal without pretilt. In this work we present an extended description of depolarizer by analyzing its electro-optic properties measured in spatial and time domains with the use of crossed polarizers and polarimetric configurations. Crossed polarizers set-up provides information on spatial and temporal changes of microscopic textures while polarimetric measurement allows to measure voltage and time dependence of degree of polarization. Three different thicknesses, i.e., 5 µm, 10 µm and 15 µm have been manufactured in order to analyze another degree of freedom for this type of depolarizer device based on a liquid crystals' material. Consideration of the light scattering capability of the cell is reported.

Keywords: depolarization; liquid crystals; Mueller matrices

1. Introduction

Depolarized light is very useful in optical measurements continuously finding application as for example spatial diffused phase element in holography [1]. Commercially used light sources are polarized or at least partially polarized. However, in many spectroscopic devices it is desirable to work with unpolarized light. Therefore, special devices should be developed to depolarize it. In general, light can be depolarized by reflection from diffused media or transmission through specially designed birefringent or scattering optical elements. Light backscattered from materials with a high roughness is a good depolarizer [2–4] and carries valuable information on optical properties of illuminated surfaces [5,6]. However, it is difficult to use such light as a source due to random direction of reflected light. This makes necessity to use the lenses' system in a detection unit because of a photodetector's limited active area.

The most popular and currently used depolarizers are designed in two ways as spatially distributed linear retarders or a composition of at least two birefringent plates having optical thickness greater than coherence length of the broad band light source. Spatial birefringence distribution has brought common forms known as Cornu or wedge types of depolarizers [7], and spectral birefringence distribution as Lyot depolarizer [8]. They are used in both, bulk optic systems [1] and optical fiber systems, as well [9].

Narrow band and laser line sources are difficult to be depolarized and both mentioned above types of optical elements are insufficient to depolarize such light sources. Therefore, an effective depolarization of such light is performed by illumination of a transparent diffused material or an optical element having spatially distributed birefringence. In the first case the transmitted light is depolarized but optical losses increase sufficiently due to the scattering effect. In the latter every particular point of an output light beam is polarized but the mean Stokes vector integrated in the cross-section of the beam shows it as depolarized light. For this reason the process is named pseudo-depolarization.

Passive depolarizers and active depolarizers based on liquid crystals (LCs) were early identified as effective devices for light depolarization. Devices available commercially and described in the literature give possibility to depolarize both narrow and broad band light sources. It is made by using scattering properties or a spatially distributed birefringence of the used LC based materials. Scattering properties of silica nanoparticles in LCs matrix [10] and cholesteric LCs in a wedge configuration [11] were presented as effective depolarizers. Spatial distribution of birefringence in LCs' polymers and different processes of LCs' polymer alignment layers were shown, as well. A commercially available depolarizer of linearly polarized light consists of stripped patterns of microretarder arrays with variable orientations of fast axes. This array is manufactured with an LC polymer [12]. Other types of spatial distributed birefringence are based on modifications of alignment layers of standard LCs' cells by using mechanical microrubbing [13], photoalignment [14], and homeotropic alignment by using surfactants [15] for selected nematic LCs.

The last configuration of the LCs was identified as an optical element with the best developing potential in designing an LC based depolarizer. Regarding the control of Vertically Aligned Nematic (VAN) cells by means of a substrate surface treatment, there are different methods for achieving better control of such device [15]. Any one of these methods has its advantages and disadvantages and what is useful for some applications may not be useful for another one. If the surface alignment layer is modified to have a given pretilt angle [16], then the LC directors will always have a predetermined reorientation upon switching by external electric field in order to find application in spatial light modulators. However, in the case of Pure VAN (PVAN) cell with zero pretilt, the local orientation of LC director will be undefined upon switching. In this case, the cell usually generates disordered birefringent medium related to undefined switching direction of molecules which produce random polarization of the transmitted light by LCs' cell without scattering. Therefore, depolarization effect may be performed and the cell cannot be used for the phase control. However, it can be used as a depolarizer which is effective for either monochromatic light or light with any spectral range.

Most of depolarizers based on LCs described above are passive optical elements. Only devices with pure homeotropic LC director configuration allow to use LC electro-optic properties in order to tune degree of polarization (DOP) level by electric field [17]. In this paper we present an extended analysis of depolarization properties of PVAN cells. The key issue to use the above concept is to have a proper spatial distribution of birefringence upon switching. To reach this goal, a biopolymer alignment layer like deoxyribonucleic acid (DNA) derivatives crosslinked with surfactant complex such as hexadecyltrimethylammonium chloride (CTMA) needs to be used [15]. When the LC material is introduced close to the surface, LC molecules align parallel to hydrophobic tails of the surfactant surrounding the DNA, thereby aligning homeotropically with the surface.

In the first stage of the study depolarization effectiveness of the PVAN cells, DOP measurements were used. This parameter strongly depends on spatial distribution of birefringence [1,17] and spectral characteristics of light sources used in the experiments [18], as well. For this type of optical element the measured DOP depends on size of the incident beam and its input state of polarization (SOP). Therefore, a narrow band red line of the stabilized He-Ne laser has been used as a light source. All manufactured PVAN cells were characterized in frame of their electro-optic properties and were analyzed in spatial domain, as well as time domain by using crossed polarizers and polarimetric set-up. Additional measurement of Mueller matrices allows to characterize dichroic and birefringence properties of the manufactured PVAN cell. Real time control of micrographic images of PAVN cell gives information about long term stability of depolarization effect of the proposed device. This type of optical element was applied to validate depolarization sensitive interferometric system [19].

2. Technology of the PVAN Cell

2.1. Preparation of Biopolymer Alignment Layer

DNA-based biopolymer was used in this research with set of experiments in order to stabilize its aligning properties. The pure DNA, as a linear and unbranched biopolymer, is soluble only in aqueous solutions which is not compatible with typically devices fabrication processes. Definitely, it is convenient to deposit thin alignment layers of DNA complexed with a suitable cationic surfactant. Such modification makes surfactant complexes of DNA soluble only in organic solvents. Additionally, this alignment surface appears to be a stable at high processing temperatures with no visible degradation of the film. The required modification of DNA with some cationic surfactant complex (see Figure 1), such as hexadecyltrimethylammonium chloride (CTMA) or dimethyldioctadecylammonium chloride (DODMAC) was done based mainly on the procedure mentioned in references [20,21]. In the aqueous solution DNA with a cationic surfactant combine through ion exchange mechanism in which the sodium ions Na$^+$ present in DNA salt back bone are replaced by surfactant groups [15] as is presented in Figure 1.

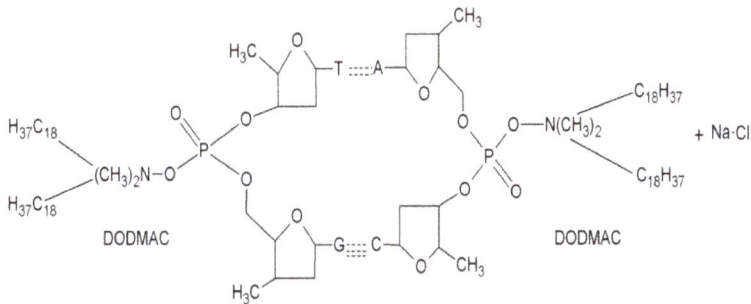

Figure 1. DNA structure with a surfactant DODMAC, where: A, T, G and C—four nitrogen-containing nucleobases and hydrogen bonds bind the nitrogenous bases of the two separate polynucleotides.

CTMA was chosen because it is a standard surfactant with a single aliphatic tail for technological applications of DNA complexes. The choice of DODMAC surfactant was motivated by its two long aliphatic tails supposed to more tightly fill the space around the DNA helix as compared to CTMA. Long alkyl chains of the cationic surfactant molecules are oriented perpendicular to the film plane and chiral DNA helices are oriented in the direction parallel to the film plane (perpendicular to the long alkyl chain) because of electrostatic attraction and thermodynamic stability [22].

The DNA conversion was performed according to the procedure in which the aqueous solution of the DNA polymer was added to an equal amount of the DODMAC aqueous solution and the precipitate was collected by filtration under vacuum and purified by rinsing with de-ionized water. DNA-DODMAC was dissolved in butanol and the solution was mixed in a glass bottle at 60 °C. Once completely dissolved, the solution was filtered through a 0.4 μm pore size syringe filter.

2.2. Manufacturing Process of the Cells

Test cells for DOP measurements were prepared using glass plates coated with conductive layer of Indium Thin Oxide (ITO) having resistivity of 20 Ohms/square. Monopixel of a 1 cm^2 area was patterned in the ITO by photolithography process. DNA-DODMAC complex was dissolved in butanol at a concentration of 3 wt. %. The solution was spin coated onto the substrates. Then, the film was baked at 80 °C for 1 hour to remove residual solvent. The two substrates were assembled using epoxy glue and uniformly separated by 5 μm, 10 μm and 15 μm thick glass spacers. All cells were filled with the experimental LCs' mixture having negative dielectric anisotropy described below.

2.3. Liquid Crystal Formulation

A high number of excellent nematic mixtures having positive dielectric anisotropy and high chemical stability have been formulated. However, very few mixtures with negative dielectric anisotropy have been developed. Commercially available stable nematic materials for VAN displays are mainly 1,2-difluorobenzene derivatives having Δn around 0.09. Efficient depolarizer based on the PVAN configuration requires sufficient retardance in the generated domains upon switching. In this case, LCs material with a relatively high birefringence will be needed, in order to achieve sufficient retardance at least $\pi + 2$ $m\pi$ for linear polarization and $\pi/2 + m\pi$ for circular polarization, for m = 0, 1, ... [15]. As the potential exploitation of this device is related to LC properties, methods of their syntheses and performance improvements related to cell manufacturing are important. The LC mixture used in this work is an experimental mixture under code name of 2050 prepared by Military University of Technology of Warsaw, Poland. This mixture has been prepared through a three- component eutectic mixture (see Table 1). To increase birefringence, we selected laterally difluoro-substituted terphenyls Compound (I) [23,24]. They have fully aromatic structure with a negative dielectric anisotropy and exhibit excellent chemical and photochemical stability. Compounds (II) and (III) are very convenient components to decrease the melting point of three ring eutectic mixtures.

Table 1. General molecular structures of compounds used to form the investigated 2050 nematic mixture and their weight %.

Components	Chemical Structure	Weight %
I	R1—⟨⟩—⟨⟩—⟨⟩—R2 (F, F) R1 and R2 = alkyl (CH_3-C_5H_{11})	34.1
II	C_3H_7—⟨⟩—⟨⟩—OCH_3	18.6
III	C_3H_7—⟨⟩—⟨⟩—OC_2H_5	47.3

The resulting mixture 2050 has negative dielectric anisotropy and exhibits the following optical parameters measured at λ = 589 nm: ordinary index n_o = 1.5006, extraordinary index n_e = 1.6273 and birefringence Δn = 0.1273. The phase sequence as a function of the temperature of the nematic mixture is Cr 0 < N <51.4 °C Iso. Temperature of phase transition between isotropic and nematic phase (clearing temperature) is around 51.4 °C.

Dielectric spectroscopy was performed to confirm the negative electric anisotropy of the proposed mixture. The procedure of this measurement is described in reference [25]. The electric anisotropy measured at room temperature (25 °C) and at the frequency of 1 kHz is about $\Delta\varepsilon$ = −0.8. This LC mixture shows a very low dielectric anisotropy since constituent compounds of the mixture do not contain strong polar groups like -CN or -NCS.

3. Electro-Optic Measurements

Polarimetric and crossed polarizers set-ups were used to study depolarization properties of manufactured PVAN cells. Both were integrated to use the same light source what was schematically presented in Figure 2. This set-up consists of spatial filter module (SF—Thorlabs), SOP generator (PSG—Thorlabs), beam splitter (BS—Thorlabs), mirror M, a pair of SOP analyzers PSA1 and PSA2 (PAX5710VIS, Thorlabs), CCD camera (DMK 72AUC02, The Imaging Source) and function generator (FG-Agilent), and personal computer (PC) as a control unit.

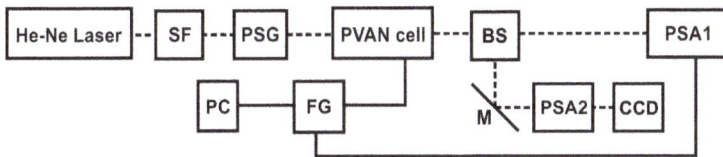

Figure 2. Integrated polarimetric and crossed polarizers' set-ups for depolarization properties characterization of a PVAN cell. SF—spatial filter module, PSG—SOP generator, BS—beam splitter, PSA1—polarimeter as a SOP analyzer, M—mirror, PSA2—SOP analyzer as a linear polarizer with vertical orientation, FG—function generator, PC—personal computer.

Laser light (He-Ne λ = 633 nm) illuminates the spatial filter module (SF) first. A Gaussian linearly polarized mode of He-Ne laser is transformed by SF to obtain light beam with homogeneous intensity and by using lenses and diaphragm the beam size is adjusted up to around 5mm of its outer diameter. Such beam passes through a SOP generator (PSG) which was assembled with a linear polarizer and a quarter waveplate. By using PSG, the manual adjustment of demanded input SOPs is possible. Light beam with selected SOP is next transmitted through the PVAN cell and polarization changes are measured by a SOP analyzer PSA1. Micrographs of the tested sample texture are taken by CCD camera after sequence reflections of a light beam from BS and M, and its passing through the second SOP analyzer PSA2. In the polarimetric part of the set-up as PSA1 the commercial PAX5710VIS polarimetric head (ThorLabs) was applied. PSG and PSA2 form the crossed polarizers' part of the set-up because PSA2 is a simple linear analyzer. Function generator (FG) drives the tested LC cell to obtain its electro-optic characteristics. PC was used to operate FG, CCD and PSA1.

Voltage changes of the FG signal allow to induce an electric field inside the LC cell to reach birefringence spatial modulation [17]. Then, measured by PSA1 Stokes vector is a spatial integration of all SOPs present in the cross-section of the transmitted beam. Since this beam carries mixed SOPs the measured Stokes vector shows this light beam as a depolarized one.

For proper validation of depolarization properties of a PVAN cell in all further experiments as a driven signal was used square waveform with a 1 kHz frequency and its voltage was changed within the range of 0–10 V with an increment of 50 mV. Additionally, thickness of the PNAN cell was taken into account and there were tested samples with thicknesses of 5 μm, 10 μm and 15 μm. In each measurement, the tested sample was illuminated first by linear horizontal (H) SOP and full polarization characterization of this device another specific input SOPs were generated, i.e., another three linearly polarized with azimuth: vertical (V), at 45° and at −45°, and two circularly polarized with right (R) and left (L) handedness.

The above measurements allow to obtain Mueller matrices for full characterization of tested samples. However, as the first validation of PVAN cells' principles of work, DOP changes for two representative input SOPs, i.e., H and R were presented in Figure 3, respectively.

The above presented DOP properties as a function of the applied voltage for LC response prove that PVAN cells are the effective depolarizers at certain values of voltages. This effectiveness has periodic character and with increasing thickness of the cell, the number of extremes increases proportionally. Number of minima doubles for circular R SOP (Figure 3b) comparing to H SOP (Figure 3a). However, DOP for linear SOP reaches lower values of minima than the circular one. This effect is related to induced by applied voltage linear birefringence of the used PVAN and is similar to results presented in [17]. As it was mentioned above, the DNA-surfactant film is expected as a promising homeotropic alignment film with rubbing-free. As result of the anchoring strength of homeotropic alignment is weak, therefore Freedericksz transition is soften to around 2 V [26].

Figure 3a,b shows that dynamics of DOP changes above this voltage increases sufficiently with LC thickness. For the sample with a thickness of 5 μm (Figure 3 green dotted line) the effectiveness of depolarization is the lowest and it has only one minimum for linear polarization and DOP is around 18% at 2.65 V (Figure 3a). For samples with thicknesses of 10 μm and 15 μm first minima appear just above the threshold voltages for both input polarizations. And for a 10 µm sample (blue dashed lines) DOP reaches its global minima for both input SOP, i.e., 4.4% at 2.1 V for horizontal and 8% at 1.95 V for right circular. In contrast for the sample with a thickness of 15 µm these minima are of 7% at 2.65 V and 11% at 3 V. These values suggest that the PVAN cell with a 10 µm thickness is the most effective among analyzed samples. Moreover, these measurement data show that PVAN cell with a 5 µm thickness is too thin to be used as an effective depolarizer and in further analysis this sample was excluded.

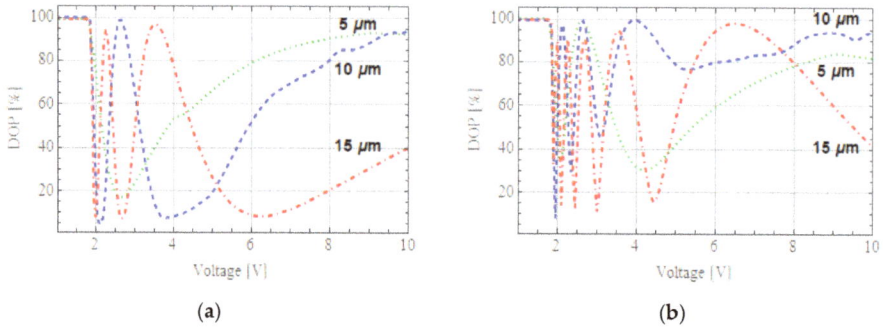

(a)

(b)

Figure 3. DOP as a function of the voltage for PVAN cells with thicknesses: 5 µm (green dotted line), 10 µm (blue dashed line) and 15 µm (red dot-dashed line) for input SOP: (a) linear H and (b) circular R.

However, an extended analysis of depolarization properties requires measurement of Mueller matrices of a PVAN cell. Mueller matrix of an LCC sample allows to characterize losses, dichroism, birefringence, and depolarization [4,27]. In this paper, the first of these parameters was excluded from consideration because above Fréedericksz transition, all manufactured samples exhibit small total losses around 0.5 dB. Due to the fact that random orientation of molecules in such type of LCC induces a random fluctuation of the refractive index, the light scattering capability of the cell has to be investigated [28]. This effect was described as a small angle light scattering in [29], and its influence on depolarization properties of the PVAN cell is discussed at the end of this paper. To obtain information about next three parameters, the polar decomposition method was used [4,27,30]. The normalized experimentally Mueller matrix M in this model is a concatenation of three matrices M_D, M_R and M_Δ which carries information about dichroism, birefringence and depolarization, respectively. This takes the following mathematical form of:

$$M = M_\Delta\, M_R M_D \tag{1}$$

Depolarization properties of optical element beads on Mueller matrix can be calculated based on Matrix M_Δ or directly from the experimental matrix M as average DOP (AvDOP) [4,27] and anisotropic depolarization degree (Add) [4]. In the paper, direct model and mentioned parameters are calculated by the following equations:

$$AvDOP = \frac{1}{4\pi} \int_0^\pi \int_{-\frac{\pi}{4}}^{\frac{\pi}{4}} DOP(\alpha, \varepsilon) \cos 2\varepsilon d\alpha d\varepsilon \tag{2a}$$

$$DOP(\alpha, \varepsilon) = \frac{\sqrt{S_1'^2(\alpha, \varepsilon) + S_2'^2(\alpha, \varepsilon) + S_3'^2(\alpha, \varepsilon)}}{S_0'(\alpha, \varepsilon)} \text{ if } S' = M\,S \text{ and } S = \begin{bmatrix} 1 \\ \cos \alpha \cos \varepsilon \\ \sin \alpha \cos \varepsilon \\ \sin \varepsilon \end{bmatrix} \tag{2b}$$

$$Add = \frac{AvDOP_{max} - AvDOP_{min}}{AvDOP_{max} + AvDOP_{min}} \tag{2c}$$

where: α and ε are azimuth and ellipticity of input SOP represented by Stokes vector S.

AvDOP is an integral of DOP calculated from output Stokes vector S' which depends on general SOP of the input Stokes vector S and Mueller matrix M of the tested optical element. Non-depolarizing optical element has AvDOP = 1 and it is 0 for totally depolarizing element. Partial depolarizing devices have intermediate values. In Figure 4a these parameters for the tested PVAN cell with a thickness of 10 μm and 15 μm were presented. Add parameter is a cumulative information about anisotropy of the average DOP calculated as a relative difference of maximal and minimal values of AvDOP. Due to the fact that DOP characteristics presented in Figure 3 have different locations of minima and maxima for linear and circular input SOPs on a voltage scale, it is expected that tested samples can be considered as at least partially anisotropic element. Therefore, it is reasonable to calculate Add parameter. In this case Add = 0 means that the depolarizer is isotropic and Add = 1 means that it is totally anisotropic. In Figure 4b Add parameters of the tested samples are presented.

Analyses of plots presented in Figure 4a show that both tested elements have relatively high values of AvDOP parameter because the minima are of around 0.35 and it is far from the measured DOP presented in Figure 3. However, information about AvDOP gives general information on how strong attention needs to be taken to apply proper voltage to a known input SOP or how to tune voltage to minimize DOP if input SOP is unknown. Unfortunately, anisotropic properties of depolarization of tested samples presented in Figure 4b in this form do not allow to apprehend if high or low value of Add is preferable or not for this device. Therefore, following the method presented in [31], the 2D map of AvDOP as a function of input SOP parameters α and ε is shown in Figure 5. Voltage values selected for these plots were taken as minima of linear and circular input SOPs, respectively.

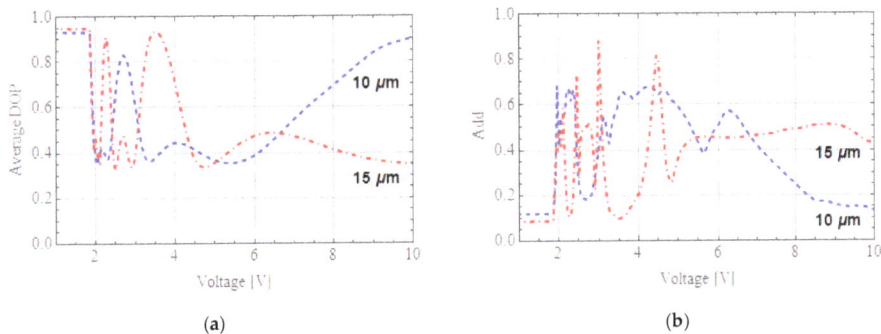

(a)   (b)

Figure 4. Calculated (**a**) AvDOP and (**b**) Add as a function of the applied voltage for PVAN cells with thicknesses of 10 μm (blue dotted line) and 15 μm (dot-dashed line). AvDOP—average DOP, Add—anisotropic depolarization degree.

The above 2D maps show zones labeled by AvDOP values. Red zones point maxima while blue zones point minima of this parameter. Proper selection of voltage allows to reach minimum DOP for a certain input SOP (Figure 3), but the above 2D map allows to access the influence of changes of input SOP on measured DOP. Thus, if the input SOP is linear H as in Figure 5a,c, for both samples it is necessary to keep SOP linear to be in or close to the DOP global minimum. Input circular polarizations [see Figure 5a,d] have higher AvDOP and their behavior in both samples is opposite to previous plots. Here minima are close to circular SOPs and changes of input SOP in the linear polarization directions locate maxima around the linear V SOP.

In the next step of the analysis the polar decomposition defined by relations (1) was used to extract information about dichroism and birefringence of the tested PVAN cell. Voltage changes of representative, cumulative parameters for mentioned above effects as diattenuation and retardance [30] are presented in Figure 6.

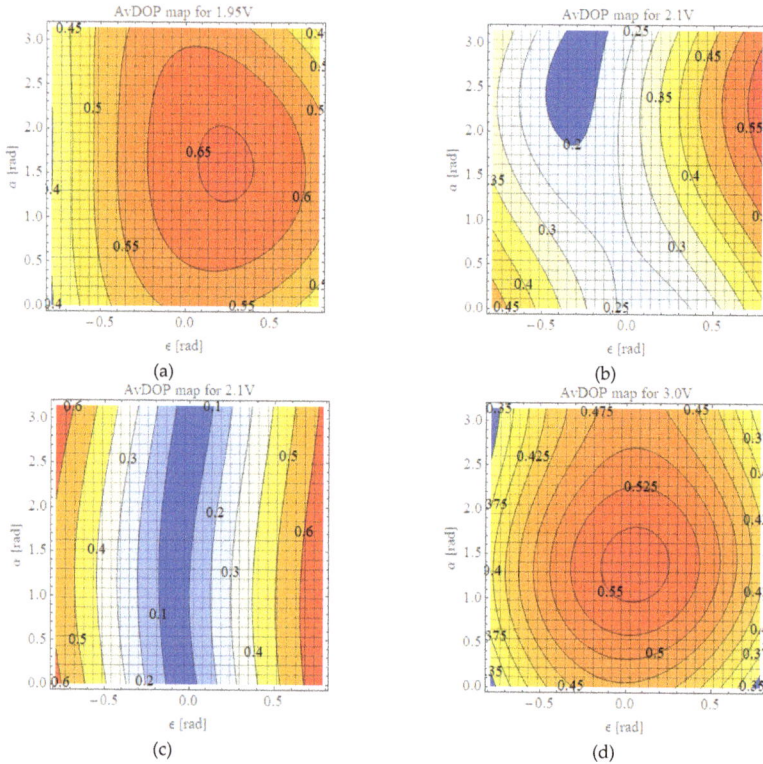

Figure 5. Calculated AvDOP maps for PVAN cells with thicknesses: 10 μm (**a**) & (**b**) and 15 μm (**c**) & (**d**) for following voltages and input SOPs (**a**) 1.95 V & circular R, (**b**) 2.1 V & linear H, (**c**) 2.1 V & linear H and (**d**) 3 V & circular R.

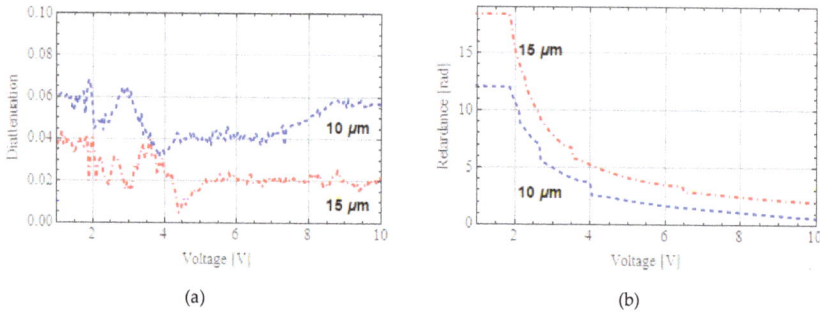

Figure 6. Calculated: (**a**) diattenuation and (**b**) retardance of tested PVAN cells with thicknesses of 10 μm (blue dotted line) and 15 μm (dot-dashed line).

Diattenuation presented in Figure 6a calculated basing on M_D matrices has very small fluctuations and in this case both tested samples are dichroism free. However, calculated total retardance (Figure 6b) based on matrices M_R includes information about cumulative birefringence of the PVAN cell. These calculations prove that depolarization properties of these devices are strongly connected with mean birefringence. Calculated birefringence [30] from retardances of both samples are of 0.116 and 0.111 for thicknesses of 10 μm and 15 μm, respectively and these values fully correspond with data mentioned in Table 1 of the previous paragraph.

Spatial and time domain electro-optic characteristics of the tested PVAN cells were characterized based on microphotographs observed under crossed polarizers' configuration of the experimental set-up from Figure 2 and were presented in Figure 7.

Microscopic textures of a conventional PVAN cell show that in this VAN nematic LC with null pretilt, upon switching by external electric field, the field causes continuous deformations of the LCs' molecules. As result of the conflict between different orientations of the molecules, topological defects are produced. Furthermore, disclinations appear where the local orientation of LC director is undefined. On the other hand, it can be supposed that the domain formation originates from the occurrence due to undefined switching direction of the molecules' director. This domains' behavior is characteristic for the formation of numerous umbilical defects induced by applied electric field to the nematic liquid crystal with negative dielectric anisotropy, confined in cells with homeotropic boundary conditions. In this case only one integer strength type of director field deformations can be formed which are regions where the in-plane component of the director rotates through $\pm 2\pi$ ($s = \pm 1$) [32], resulting in spatial distribution of birefringence in a PVAN cell under voltage action (see Figure 7). These defects are important in practical applications, such as depolarization of polarized light. Moreover, we have observed that the morphology of such textures changes in time due to umbilical defect annihilation over time where defects of opposite sign and equal strength attract each other and annihilate, thus reduce the number of observed defects with time [33]. Similar results have been observed in our experiment as it is shown in the sequences of Figure 7a to Figure 7b and to Figure 7c for 10 μm thick sample and next Figure 7d–f for 15 μm thick cell. The annihilation dynamics of nematic umbilical defects, induced by electric field application to homeotropically oriented liquid crystal samples of negative dielectric anisotropy, were experimentally investigated in [34].

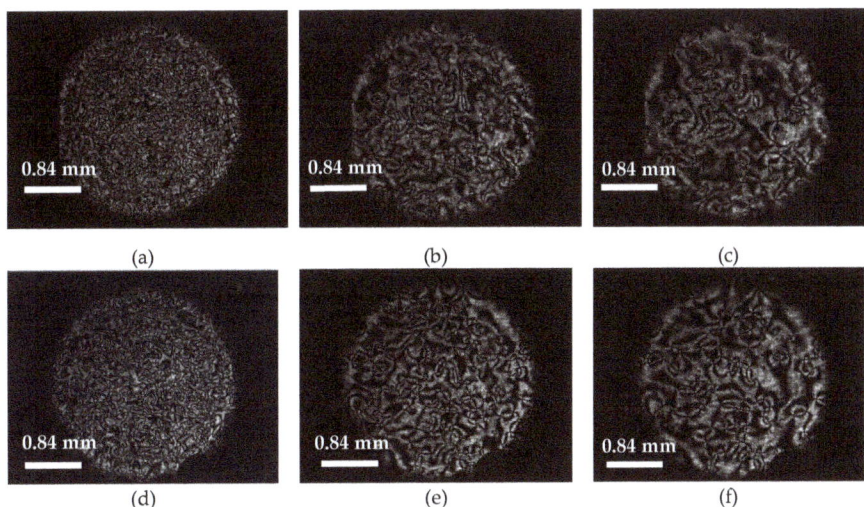

Figure 7. Microphotographs of 2050 LC domains on the DODA biofilm observed under crossed polarizers for input linear horizontal SOP and under voltage of 2.15 V applied to the cell in three time intervals of the samples with thickness of 10 μm after (**a**) 25 s, (**b**) 150 s (**c**) 300 s and 15 μm after (**d**) 25 s, (**e**) 150 s, (**f**) 300 s.

Figure 8 shows time evolution of the measured DOP at different applied voltages. Since the azimuthal tilting of the LC molecules is not defined, the measurement of DOP in 10 μm and 15 μm cells shows that it takes more than 200 s for the first case and 25s for the second to reach the stable DOP through the reorientation process. Similar results have been observed previously [35].

To study the light scattering capability of the cells with thickness of 5 μm, 10 μm and 15 μm, Fourier transform of scattered light has been adopted. This method has been developed as a sensitive one for studying this effect from inhomogeneous and dynamic structures [36]. The samples were illuminated using a parallel beam of He-Ne laser and placed behind the lens at a certain distance from the focal plane. We applied to each sample a voltage corresponding to minimum (DOP); V = 2.65 V, 2.15 V and 2.15 V in the case of 5 µm, 10 µm and 15 µm, respectively. We recorded the images of the scattered beams that are detected by CCD camera placed at Fourier plan after different times' lag detection. Figure 9 shows images of spectral intensities' distributions of the focused beams after switching ON of the cells with voltages corresponding to the DOP minimum. The small angular scattering distribution has been detected in Fourier plane where the beam profile differs from the focused condition in OFF state (Figure 9a,e,i) showing beam spot spreading. As results of a spatial and temporal changes of refractive index distribution due to annihilation of umbilical defects, a spatial reshaping of the beam spreading changes with time. Significant changes of the beam reshaping with time are observed in thicker cells (10 and 15 µm). However, in a thin cell of 5 µm, no significant changes in light spreading is observed. Correlation of scattering properties of the studied cells with their dynamic change of spatial refractive index distribution, thickness effect and temperature dependence will be a subject of a further study.

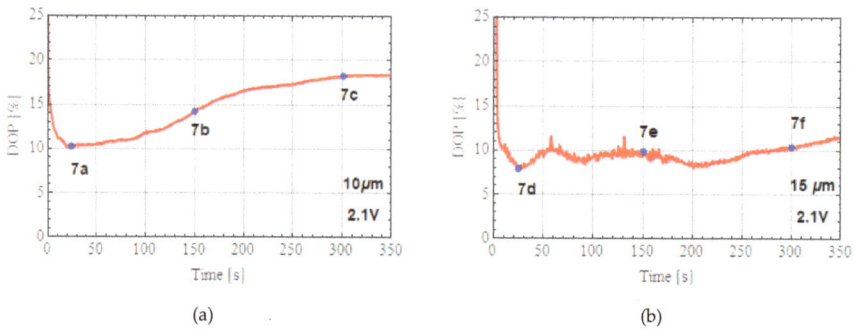

(a) (b)

Figure 8. Time evolution of measured DOP at different applied voltages to PVAN cells with thickness of: (**a**) 10 µm and (**b**) 15 µm. The SOP of the incident light is linearly H polarized.

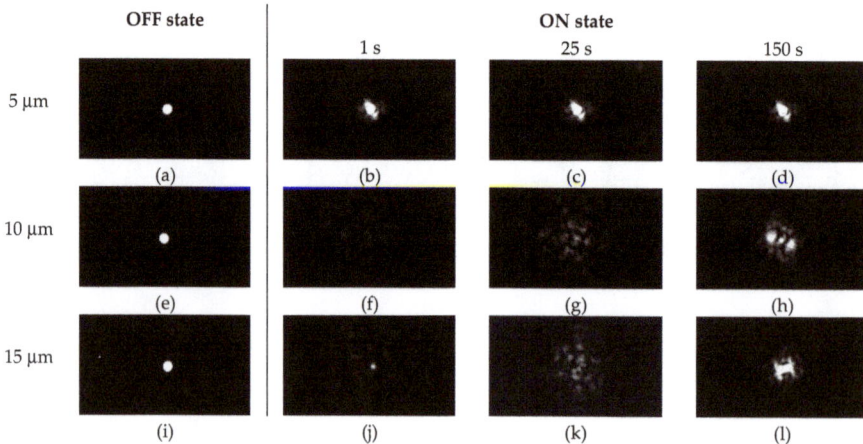

Figure 9. Spectral intensity distribution of a focused beam transmitted by cells of thickness of 5 µm, 10 µm and 15 µm when the cells are driven with the voltage of V = 2.65 V, 2.15 V and 2.15 V, respectively. (**a,e,i**) represent the focused condition at V = 0. The images of scattered beam are recorded by a CCD camera placed at Fourier plan after different times' lag detection. The SOP of the incident light is linear H polarized.

4. Conclusions

Depolarization of a monochromatic laser beam based on a pure vertically aligned nematic LC has been proposed. The alignment properties of DNA-based biopolymer were adopted to produce PVAN in order to design a high quality active pseudo-depolarizer. One of electro-optical properties of the tested PVAN cell is the ability to control the DOP level using relatively low voltages. Depolarization level of this device depends strongly on input SOP and generally has the anisotropic character. Polar decomposition method applied to experimental Mueller matrices shows that for tested devices voltage applied to the cell induces birefringence domains in addition the calculated mean values of retardance are at the level of 0.11. This value is in good correspondence with data from the refractometer measurement of the used LC. Furthermore, they are dichroism free. The small angular scattering distribution has been detected at Fourier plane after switching ON of the cells with voltage corresponding to the DOP minimum. As a result, a spatial reshaping of the beam spreading is changing with time due to annihilation of the umbilical defect.

Author Contributions: P.M. and N.B. analyzed measurement data and prepared manuscript, A.S. prepared and described the biopolymer of alignment layer, A.K. measured PVAN cells, R.W. manufactured PVAN cells, K.G. formulated the LC crystal mixture, L.R.J. analyzed and interpreted measurement data.

Funding: This work was supported by the Ministry of National Defense by research grants: GBMON/13-995/2018/WAT and Ministry of Higher Education as a statutory activity project PBS -23 - 898.

Acknowledgments: We would like to thank Anna Pakuła and Tomasz Jankowski from Virtual Reality Techniques Division of the Warsaw University of Technology for the scattering measurement.

Conflicts of Interest: The authors declare no conflict of interest.

References

1. Takeda, M.; Wang, W.; Naik, D.N.; Singh, R.K. Spatial Statistical Optics and Spatial Correlation Holography: A Review. *Opt. Rev.* **2014**, *21*, 849–861. [CrossRef]
2. Le Roy-Brehonnet, F.; Le Jeune, B. Utilization of Mueller matrix formalism to obtain optical targets depolarization and polarization properties. *Prog. Quant. Electr.* **1997**, *21*, 109–151. [CrossRef]
3. Le Roy-Brehonnet, F.; Le Jeune, B.; Gerligand, P.Y.; Cariou, J.; Lotrian, J. Analysis of depolarizing optical targets by Mueller matrix formalism. *Pure. Appl. Opt.* **1997**, *6*, 385–404. [CrossRef]
4. Boulvert, F.; Le Brun, G.; Le Jeune, B.; Cariou, J.; Martin, L. Decomposition algorithm of an experimental Mueller matrix. *Opt. Commun.* **2009**, *282*, 692–704. [CrossRef]
5. Firdous, S.; Anwar, S. Noninvasive optical diagnostic of breast cancer using depolarization of light. *Optik* **2016**, *127*, 3035–3038. [CrossRef]
6. Pierangelo, A.; Benali, A.; Antonelli, M.-R.; Novikova, T.; Validire, P.; Gayet, B.; De Martin, A. Ex-vivo characterization of human colon cancer by Mueller polarimetric imaging. *Opt. Express* **2011**, *19*, 1582–1593. [CrossRef] [PubMed]
7. Ge, J.H.; Chen, Z.; Chen, Y.F.; Chen, C.Y.; Zhai, Y.Z.; Zhang, J.; Sui, Z.; Lin, H.H.; Wang, J.J.; Deng, Q.H. Optimized Design of Parameters for Wedge-Crystal Depolarizer. *App. Mech. Mater.* **2012**, *110–116*, 3351–3357. [CrossRef]
8. Burns, W.K. Degree of polarization in the Lyot depolarizer. *J. Lightwave Technol.* **1983**, *1*, 475–479. [CrossRef]
9. Jaroszewicz, L.R.; Krajewski, Z.; Kowalski, H.; Mazur, G.; Zinówko, P.; Kowalski, J. AFORS autonomous fibre-optic rotational seismograph: Design and application. *Acta Geophys.* **2011**, *59*, 578–596. [CrossRef]
10. Diorio, N.J., Jr.; Fisch, M.R.; West, J.L. Filled liquid crystal depolarizers. *J. Appl. Phys.* **2001**, *90*, 3675–3678. [CrossRef]
11. Zhang, D.; Luo, F.; Luo, Y.; Li, J.; Liu, C.; Liu, H.; Shen, Z.; Wang, W. Cholesteric liquid crystal Depolarizer. *Opt. Eng.* **2007**, *46*, 070504. [CrossRef]
12. ThorLabs, Inc. Available online: https://www.thorlabs.com/newgrouppage9.cfm?objectgroup_id=8043 (accessed on 26 July 2019).
13. Honma, M.; Nose, T. Liquid-crystal depolarizer consisting of randomly aligned hybrid orientation domains. *Appl. Opt.* **2004**, *43*, 4667–4671. [CrossRef] [PubMed]

14. Wei, B.-Y.; Chen, P.; Ge, S.-J.; Zhang, L.-C.; Hu, W.; Lu, Y.-Q. Liquid crystal depolarizer based on photoalignment technology. *Photon. Res.* **2016**, *4*, 70–73. [CrossRef]
15. Spadło, A.; Bennis, N.; Węgłowski, R.; Węgłowska, D.; Czupryński, K.; Otón, J.M. Biopolymer as alignment layer for liquid crystal mixtures. *Mol. Cryst. Liq. Cryst.* **2017**, *657*, 56–65. [CrossRef]
16. Marino, A.; Santamato, E.; Bennis, N.; Quintana, X.; Otón, J.M.; Tkachenko, V.; Abbate, G. Ellipsometric Study of Vertically Aligned Nematic Liquid Crystals. *App. Phys. Lett.* **2009**, *94*, 013508. [CrossRef]
17. Vena, C.; Massarelli, R.; Carbone, F.; Versace, C. An approach to a model disordered birefringence medium for light depolarization applied to a liquid crystal device. *J. Opt.* **2014**, *16*, 16065705. [CrossRef]
18. Sakai, J.; Machida, S.; Kimura, T. Degree of polarization in anisotropic single-mode optical fibers: Theory. *J. Quant. Electro.* **1982**, *18*, 488–495. [CrossRef]
19. Kalbarczyk, A.; Jaroszewicz, L.R.; Bennis, N.; Chruściel, M.; Marć, P. The Young Interferometer as an Optical System for a Variable Depolarizer Characterization. *Sensors* **2019**, *19*, 3037. [CrossRef]
20. Heckman, E.M.; Hagen, J.A.; Yaney, P.P.; Grote, J.G.; Hopkins, F.K. Processing techniques for deoxyribonucleic acid: Biopolymer for photonics applications. *Appl. Phys. Lett.* **2005**, *87*, 211115. [CrossRef]
21. Wang, L.; Yoshida, J.; Ogata, N.; Sasaki, S.; Kajiyama, T. Self-Assembled Supramolecular Films Derived from Marine Deoxyribonucleic Acid (DNA)–Cationic Surfactant Complexes: Large-Scale Preparation and Optical and Thermal Properties. *Chem. Mater.* **2001**, *13*, 1273–1281. [CrossRef]
22. Grote, J.G.; Diggs, D.E.; Nelson, R.L.; Zetts, J.S.; Hopkins, F.K.; Ogata, N.; Hagen, J.A.; Heckman, H.; Yaney, P.P.; Stone, M.O.; et al. DNA Photonics [Deoxyribonucleic Acid]. *Mol. Cryst. Liq. Cryst.* **2005**, *426*, 3–17. [CrossRef]
23. Kula, P.; Spadło, A.; Dziaduszek, J.; Filipowicz, M.; Dąbrowski, R.; Czub, J.; Urban, S. Mesomorphic, dielectric and optical properties of fluorosubstituted biphenyls, terphenyls, and quaterphenyls. *Opto-Electron. Rev.* **2008**, *16*, 379–385. [CrossRef]
24. Urban, S.; Kula, P.; Spadło, A.; Geppi, M.; Marini, A. Dielectric properties of selected laterally fluorosubstituted 4,4''-dialkyl, dialkoxy and alkyl-alkoxy [1:1';4':1''] terphenyls. *Liq. Cryst.* **2010**, *37*, 1321–1330. [CrossRef]
25. Perkowski, P. Numerical elimination methods of ITO cell contribution to dielectric spectra of ferroelectric liquid crystals. *Opto-Electron. Rev.* **2011**, *17*, 176–182. [CrossRef]
26. Seo, D.-S. Generation of pretilt angle in NLC and EO characteristics of transcription-aligned TN-LCD fabricated by transcription alignment on polyimide surfaces. *Liq. Cryst.* **1999**, *26*, 397–400. [CrossRef]
27. Goldstein, D. *Polarized Light*, 2nd ed.; Marcel Dekker: New York, NY, USA, 2003; pp. 533–557.
28. Lorenz, A.; Omairat, F.; Braun, L.; Kolosova, V. Nematic copolymer network LCs for swift continuous phase modulation and opaque scattering states. *Mol. Cryst. Liq. Cryst.* **2017**, *646*, 220–225. [CrossRef]
29. Hashimoto, T.; Nakai, A.; Shiwaku, T.; Hasegawa, H.; Rojstaczer, S.; Stein, R.S. Small-Angle Light Scattering from Nematic Liquid Crystals:Fluctuations of Director Field Due to Many-Body Interactions of Disclinations. *Macromolecules* **1989**, *22*, 422–429. [CrossRef]
30. Kula, P.; Bennis, N.; Marć, P.; Harmata, P.; Gacioch, K.; Morawiak, P.; Jaroszewicz, L.R. Perdeuterated liquid crystals for near infrared applications. *Opt. Mat.* **2016**, *60*, 209–213. [CrossRef]
31. Chipman, R.A. Depolarization in the Mueller Calculus. *Proc. SPIE* **2003**, *5158*, 184–192.
32. Dierking, I.; Ravnik, M.; Lark, E.; Healey, J.; Alexander, G.P.; Yeomans, J.M. Anisotropy in the annihilation dynamics of umbilic defects in nematic liquid crystals. *Phys. Rev. E* **2012**, *85*, 021703. [CrossRef]
33. Chuang, I.; Durrer, R.; Turok, N.; Yurke, B. Cosmology in the laboratory: Defect dynamics in liquid crystals. *Science* **1991**, *251*, 1336–1342. [CrossRef] [PubMed]
34. Dierking, I.; Marshall, O.; Wright, J.; Bulleid, N. Annihilation dynamics of umbilical defects in nematic liquid crystals under applied electric fields. *Phys. Rev. E* **2005**, *71*, 061709. [CrossRef] [PubMed]
35. Vena, C.; Versace, C.; Strangi, G.; D'Elia, S.; Bartolino, R. Light depolarization effects during the Fréedericksz transition in nematic liquid crystals. *Opt. Express* **2007**, *15*, 17063–17071. [CrossRef] [PubMed]
36. Ding, H.; Wang, Z.; Nguyen, F.; Boppart, S.A.; Popescu, G. Fourier Transform Light Scattering of Inhomogeneous and Dynamic Structures. *Phys. Rev. Lett.* **2008**, *101*, 238102. [CrossRef] [PubMed]

crystals

MDPI

Article

Research on Optical Properties of Tapered Optical Fibers with Liquid Crystal Cladding Doped with Gold Nanoparticles

Joanna E. Moś [1,*], Joanna Korec [1], Karol A. Stasiewicz [1], Bartłomiej Jankiewicz [2], Bartosz Bartosewicz [2] and Leszek R. Jaroszewicz [1]

[1]   Institute of Technical Physics, Military University of Technology, 00-908 Warsaw, Poland; joanna.korec@wat.edu.pl (J.K.); karol.stasiewicz@wat.edu.pl (K.A.S.); leszek.jaroszewicz@wat.edu.pl (L.R.J.)
[2]   Institute of Optoelectronics Military University of Technology, 00-908 Warsaw, Poland; bartlomiej.jankiewicz@wat.edu.pl (B.J.); bartosz.bartosewicz@wat.edu.pl (B.B.)
*   Correspondence: joanna.mos@wat.edu.pl; Tel.: +48-261-837-001

Received: 15 May 2019; Accepted: 12 June 2019; Published: 14 June 2019

Abstract: This paper presents results obtained for biconical tapered fibers surrounded/immersed in liquid crystal mixtures. The phenomenon of light propagating in the whole structure of a tapered fiber allows the creation of a sensor where the tapered region represents a core whereas the surrounding medium becomes a cladding. Created devices are very sensitive to changing refractive index value in a surrounding medium caused by modifying external environmental parameters like temperature, electric or magnetic field. For this reason, the properties of materials used as cladding should be easily modified. In this investigation, cells have been filled with two different nematic liquid crystals given as 1550* and 6CHBT (4-(trans-4-n-hexylcyclohexyl) isothiocyanatobenzoate), as well as with the same mixtures doped with 0.1 wt% gold nanoparticles (AuNPs). Optical spectrum analysis for the wavelength range of 550–1150 nm and time-courses performed for a wavelength of 846 nm at the temperature range of T = 25–40 °C were provided. For all investigations, a steering voltage in the range of 0–200 V which allows establishing the dopes' influence on transmitted power and time response at different temperatures was applied.

Keywords: liquid crystal; gold nanoparticles; optical fiber device; tapered optical fiber

1. Introduction

Due to the rapid development of optical fiber technologies, we reach limits related to the use of materials or arrangements to measure different factors. We are looking for the possibility of using a "new material" to get an improvement in devices' parameters or to increase its sensitivity. For many years, scientists have been looking for and using combinations of different materials, using specific properties of each of them [1–3]. As a result, modified materials have been produced with new properties called hybrids. One of the interesting technological connections is the use of the fiber optic and liquid crystal technology [4,5]. This hybrid combination allows us to create an active, functional element consisting of a passive element—an optical fiber and an active liquid crystal (LC) medium, which properties can be controlled through electric field and temperature [6]. An important aspect is to find a common feature for these materials and their mode of interaction. In both cases, the common feature is the refractive index parameter of glass and LC, respectively. Use of optical fiber was related to the method of light transmission. Propagation of light in the fiber is associated with an appropriate selection of materials forming the cladding and the core with an appropriate refractive index n_{core}, $n_{cladding}$ filling the Maxwell equation/condition, as well as Snell law. The light beam can be propagated in two ways: based on phenomena of a total internal reflection (TIR) (in classic fibers) [7] or the principle of photonics' bandgap (in photonic crystal fibers, PCF) [8]. The change in the refractive index of materials caused modified properties of the propagated light in the form of modes. The value of the normalized frequency V determined the number of modes propagating in the fiber and the losses. This parameter is described as [9]:

$$V^2 = a\frac{2\pi}{\lambda} \sqrt{n_{core}^2 - n_{cladding}^2} \tag{1}$$

From the above formula, it can be concluded that the number of modes propagating in the fiber is also influenced by the wavelength λ and the diameter of the core a. This parameter is important when we want to use a tapered optical fiber [10–12] or long period gratings (LPG) [13,14] as an element of optical devices. LCs are built of elongated molecules. Such construction allows determination of the anisotropic properties of LC, characterized by ordinary n_o and extraordinary n_e reflective indices. Depending on the light propagation direction, a different value of the refractive index will be detected. Through an appropriate electrical field (higher than threshold voltage) or temperature (changes in structure), we can change the reflective index by reorientation of LC position. While combining these materials, we notice that the optical fiber will be sensitive to changes in the LC refractive index [15]. Considering the diversity of optical fibers and LC types, there are many combinations possible. Many works have been written about fulfilling the air holes of a PCF with LC which causes a change of effective refractive index of the cladding structure and, hence the change of the propagation properties [16,17]. LCs can also be used externally by constructing a liquid crystal cell (LCC) with a tapered optical fiber or LPG placed between the cell's plates [13,14]. It is important to choose an appropriate LC with its refractive index, transition temperature to the isotropic state, dielectric permittivity, structure, composition, etc. Selection of these parameters affects the parameters of later devices, e.g., time of relaxation, threshold voltage, etc. As a result, we obtain interesting devices' sensors for single factors, as well as optical filters for a chosen wavelength [13,18]. The combination of these materials provides many benefits: optical fibers can be used over long distances, device miniaturization, device tunability in a different wave range and manufacturing low costs.

The use of a tapered optical fiber and LCs has been known for many years [19,20]. It was possible to control LCs, but they were characterized by high threshold voltages, as well as by large thicknesses of LCCs. Nowadays, the newly developed technology associated with the parameters' improvement is used as a novelty—use of appropriate spacers close to the taper waist diameter, use of the LC optimal type, or use of a LC mixture with other materials such as nanoparticles [1–5]. The research was carried out on the influence of doping with gold nanoparticles on LCC properties. As a result, improvements were made to such parameters as decreased threshold voltage and response time of the LCC [3]. Our research team has investigated how nanoparticles with LC affect light propagation in the optical fiber and, more specifically, in a tapered optical fiber, as well as how to elaborate a new kind of device working in a different wave range. In the research LCs named 1550 and 6CHBT (4-(trans-4-n-hexylcyclohexyl) isothiocyanatobenzoate) were applied—well known from filling air holes in PCF fibers [21]. Novelty of these works is connected with liquid crystal doping nanoparticles and applying them as a cladding with a possibility of refractive index steering. Previous literature on the use of a liquid crystal mixture with nanoparticles as a cladding of the optical fiber taper was not found. For this reason, the presented results are unique. Tapers were manufactured from a standard telecommunication single-mode fiber (SMF) with a cut-off wavelength of 1260 nm—the most popular and low-cost material produced by Corning®. Thanks to the tests carried out on cells with pure LC and with a mixture of gold nanoparticles with LC, we can compare the results of adding nanoparticles to light propagation under voltage and temperature control.

2. Materials and Methods

2.1. Propagation of Light in Tapered Optical Fiber with and without Liquid Crystal Cladding

An optical fiber consists of two concentrically arranged dielectrics: the core and the cladding. The core is characterized by a higher refractive index than the cladding [8]. It determines light propagation in that structure based on a total internal reflection (Figure 1). The light is propagated in the form of modes in the core and a small part of energy penetrating the cladding which is named the evanescent field. During the technological process consisting of heating a certain part of the fiber and pulling it out, a tapered optical fiber can be created. The result of a fiber's tapered structure is change in the core and cladding diameters (Figure 2). As a result, the field of the basic mode occupies an increasingly larger cross-section of the cladding until it reaches the point where it is guided in its entire cross-section. As a result, the field of the basic mode occupies an increasingly larger cross-section of the cladding until it reaches the point where it is guided in its entire cross-section.

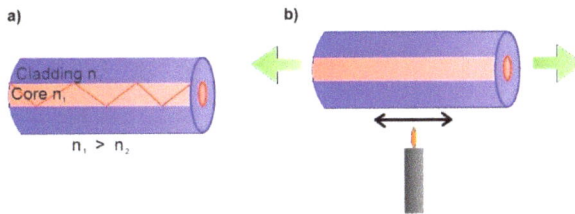

Figure 1. (**a**) Light propagation in optical fiber and (**b**) process of heating and elongation of the optical fiber.

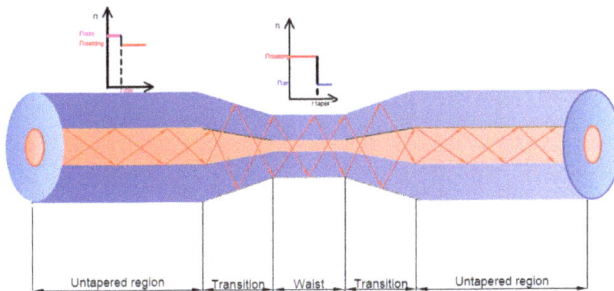

Figure 2. The scheme of light propagation in a tapered optical fiber—changes of boundary conditions.

The change of boundary conditions is also directly related to the change of optical fiber dimensions. The light is not propagated at the core/cladding interface, but the air/cladding interface. The approximations used in the mode solutions for a standard optical fiber, where the difference in refractive indices is small (lower than 1%), are not applicable. In this case, in the tapered optical fiber, the light beam stays to be exposed to influence by the fiber surroundings—the light beam is propagating in a whole structure (as a core) and the surroundings become a cladding as is shown in Figure 2 [8,9].

Such an approach allows us to use additional material in a tapered area especially in the taper waist region (Figure 3). In this region, the optical fiber will be sensitive to external changes of the reflective index of the material that surrounds the fiber. Such connections with various materials have been widely used to build sensors, filters, and amplifiers [22,23].

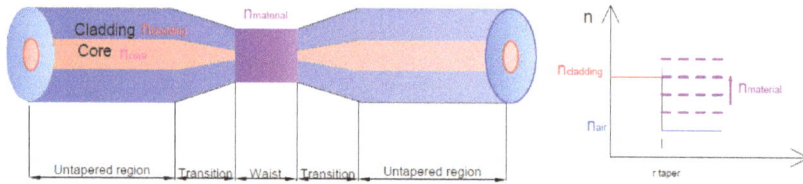

Figure 3. The tapered optical fiber with extra material.

The reason for taking liquid crystals as a research object is that they are anisotropic materials possessing variable refractive index depending on the molecules' orientation. It should also be mentioned that in many cases LCs have a higher refractive index than an optical fiber (propagation changes from total internal reflection to the band gap propagation or a mix of them for different wavelengths) which makes them most interesting materials. Additionally, LCs' substrates create a resonance device which causes the situation that some part of the light is reflected from the glass with alignment layers and can be coupled back to the fiber taper. In a standard single-mode fiber (SMF) the basic mode is propagated. When we taper such a fiber, boundary conditions change and we can see many modes propagating especially in the taper waist area. An attitude mode begins to couple with cladding modes so that the light beam is modified. The light starts to leak from the taper's structure to the liquid crystal cladding as shown in Figure 4. In LC, light beams are limited by the glass with alignment layers as it was mentioned earlier. Depending on the used voltage, the angle of LC scattering increases [see Figure 4b] [15,24].

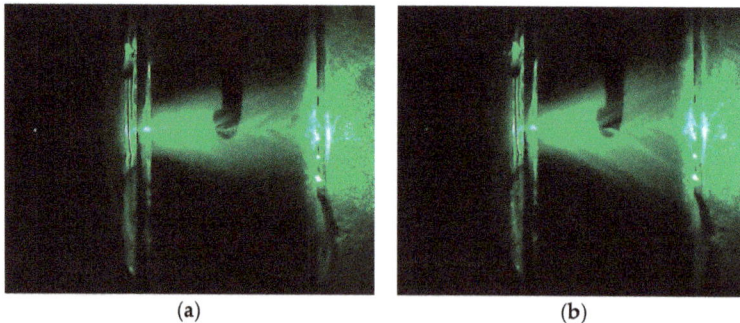

Figure 4. Light propagation in the liquid crystal cell with the tapered fiber: (**a**) U = 0 V; (**b**) U = 200 V.

2.2. Materials

In this paper were investigated two nematic LCs: 6CHBT (4-(trans-4-n-hexylcyclohexyl) isothiocyanatobenzoate) and 1550*, also the same mixtures doped with 0.1 wt% AuNPs. Table 1 presents the main properties of the pure LCs used in this paper. 6CHBT LC is a low melting LC and compared to the 1550*, 6CHBT LC contains only one mixture component. Also, there is a significant difference between the refractive indices of both LCs. As presented in Table 1, 6CHBT has a much higher refractive index than 1550* and is higher than silica glass in room temperature. According to the literature, in some cases, the light can be propagated in the core with a lower refractive index than the effective cladding refractive index [23,25].

Table 1. Properties of the used liquid crystals (LCs): 6CHBT (4-(trans-4-n-hexylcyclohexyl) isothiocyanat obenzoate) and 1550* [21,26].

LC	Structure		n_e (589 nm)	n_o (589 nm)	Δn (589 nm)	$\varepsilon_\perp$ (1 kHz)	$\varepsilon_{\parallel}$ (1 kHz)	T_{ISO} [°C]
6CHBT		C_6H_{13} — NCS	1.672	1.518	0.154	4.3	12.0	43.0
1550*	23.93%	C_3H_7 — OCOO C_2H_5	1.528	1.462	0.066	2.15	5.25	79.8
	18.24 %	C_5H_{11} — OCOO CH_3						
	34.55 %	C_5H_{11} — OCOO C_2H_5						
	23.28 %	C_3H_7 — CN						

Basing on the results from [27], the authors decided to mix pure LCs with 0.1 wt% AuNPs. Preparation of nanoparticle (NP)-doped LCs has been carried out in the same manner as described in detail elsewhere [28]. Briefly, the colloidal suspension of Au NPs in an organic solvent was prepared using the Brust–Schiffrin method [29]. However, instead of toluene, we have used chloroform. After synthesis and purification, we have obtained suspension of Au NPs in chloroform with a concentration of 1.2 mg/mL. Au NPs had diameters in the range of 1–3 nm. To obtain NPs-doped LCs with various Au NPs content, we have mixed various volumes of the colloidal suspension of AuNPs with liquid crystal and evaporated chloroform at 50 °C for 48 h.

2.3. Technology

This section presents the manufacturing of an LCC with the tapered optical fiber. The FOTET (fiber optic taper element technology) system (see Figure 5) was used for tapered fibers' manufacturing. FOTET is a dedicated device for preparing different optical fiber elements, e.g., couples, tapers or isolators. The main advantage of this technique is the possibility to modify different types of optical fibers, e.g., single mode, multimode, as well as photonic crystal fibers [30] and to obtain different shapes of tapers: adiabatic, as well as non-adiabatic.

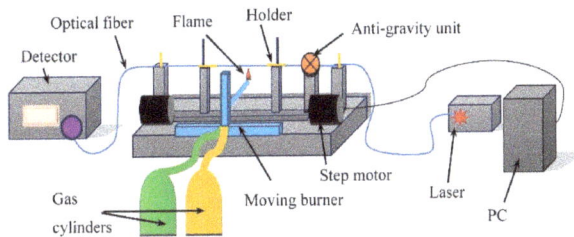

Figure 5. Scheme of the FOTET (fiber optic taper element technology) system's arrangement of the tapered fiber manufacturing.

Tapering is performed by stretching the optical fiber above the flame from a low-pressure burner powered with a propane-butane-oxygen gas mixture. An appropriate proportion of this mixture and distance between flame and fiber allow for controlling flame temperature and, connected with it, drawing velocity. In this process, the temperature is most important because too high a temperature causes melting of the optical fiber and too low a temperature is not enough to soften the fiber, which results in stopping the stretching process. Appropriate temperature is the softening temperature of glass (around 1200 °C). A pure mixture of propane–butane and oxygen gases allows to obtain a homogenous and adequate temperature distribution. The velocity of fiber elongation is controlled by software which is based on a measurement of the transmitted optical signal magnitude. Additionally, the special anti-gravity unit is used because during the elongation process the fiber position is reduced. The signal increase in the anti-gravity system is an impulse for changing drawing velocity, as well as lowering the burner position [11,12]. Additionally, by selection of a proper flame, we can make a taper with a very small dimension of the taper waist. The main advantage of preparing tapered fibers on the FOTET are full control elongation parameters like elongation length, burner movement, and its distance from the fiber or step motors velocity, taper waist region diameter. Additionally, during elongation, changing optical power is constantly monitored by using a laser and a photodetector connected to the opposite ends of the fiber. Thus, attenuation after the finished tapering process can be easily calculated.

Prepared tapers used for manufacturing devices are characterized by low losses below $\alpha = 0.2$ dB, elongation L = 20.20 ± 0.05 mm, and diameter $\varphi = 14.50 ± 0.50$ μm for a wavelength of 1550 nm which corresponds to the single mode work of the used single mode fiber.

In the next stage, the tapered fiber LCCs were prepared. The initial arrangement of molecules is achieved by using a photoalignment poly(estermide) layer on the glass substrate covered with ITO (Indium tin oxide) (further called electrode) [31]. The obtained tapered fiber is placed on the electrode as close as possible but without touching it. The position is obtained with control of a CCD (Charge Coupled Device) camera with special objective lenses. Any contact between the tapered fiber and the electrode causes radiating out optical power and increasing losses. To maintain the same distance between electrodes, spacers with a diameter of $\varphi = 40$ μm were used. In this paper, the perpendicular arrangements of taper and electrodes were investigated. In this case, the top and bottom electrodes have a perpendicular direction of the alignment layer with respect to the fiber/taper axis. Prepared cells are filled with LCs right before the measurement to ensure conglomeration on the NP. Figure 6 presents the scheme and cross-section of the prepared cell.

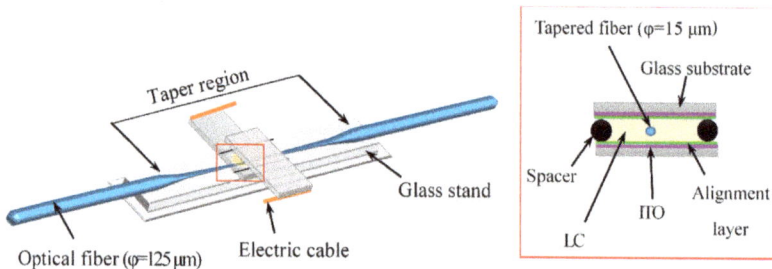

Figure 6. Scheme of the liquid crystal cell with a tapered fiber.

In general, the different refractive index of a waveguide cladding (LC around tapered fiber) depends on the orientation of n-director molecules in the external electric field. However, around the taper, an area of disoriented LC molecules is formed which is attached to the fiber as it is shown in Figure 7 [32]. This part of the LC cannot be reoriented even with a high voltage, which is connected to adhesive strength between glass and LC molecules. Nevertheless, its diameter is smaller than the wavelength. Hence, it will not be under investigation in this paper. It also should be noticed that in this investigation only the medium refractive index, which can be described as the medium molecules' director, is taken under consideration. In the doped LC topological defects are induced around nanoparticles (See Figure 7). Even small amounts of NPs significantly influence the properties of bulk material due to intermixing NP effects on alignment layers and bulk properties [33]. Additionally, we can predict that in a temperature change process NPs will be new small centers of expanding temperature.

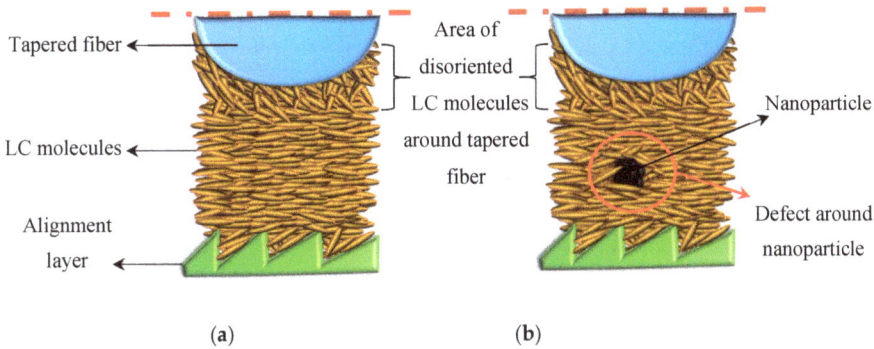

Figure 7. Scheme of molecules' orientation around the taper and in the whole volume of the LC cell: (**a**) pure LC; (**b**) LC doped with nanoparticles.

Results obtained show the positive and negative influence of dopants. Nanoparticles' behaviour in these liquid crystals can be explained by the difference between physical and chemical properties of 6CHBT and 1550*. In contrast to 6CHBT, 1550* is a mixture of four components with much higher molar mass than the first one. The differences in measured properties are mainly connected to the concentration and size of nano-molecules in the solution.

3. Results

Applied voltage influence on changes in spectral characteristics was investigated. Figure 8 shows the scheme of an optical system for measuring changes in spectral characteristics depending on voltage. For the measurement a high-power supercontinuum laser (SuperK EXTREME, NKT Photonics, Southampton, UK) (since light operates in the wavelength range of 400–2400 nm), an optical spectrum analyser with the spectral range of 350–1200 nm, (AQ6373, Yokogawa, Tokyo, Japan), signal generator (DG1022, RIGOL Beijing; China), and voltage amplifier (A400D, FLC Electronics, Molndal, Sweden) were used.

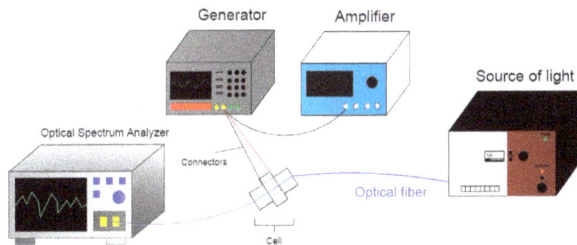

Figure 8. Scheme of the optical system for measuring changes in spectral characteristics depending on voltage.

Figure 9 presents the spectral characteristics for a cell filled with 1550* and 6CHBT LC for pure and doped NPs with voltage steering without modulation in the range of 0–200V at room temperature. For 1550* LC, a reduction of power transmission in almost whole wave range can be observed. Steering by an electric field causes an increase of power light propagated in the optical fiber device as can be seen in Figure 9a,b. Much better transmission is observed for 1550* doped Au NPs. In all wave ranges, we observe smaller losses than for pure LC. Together with applying voltage power in the full range, optical power is increasing and, additionally, broadening the wave range can be observed in comparison to the pure LC. For pure 6CHBT we observe smaller attenuation of optical power in whole range than for pure 1550* mixture. Furthermore, changes in optical power regarding the applied electric field are insignificant. For 6CHBT with nanoparticles, interference between modes are observed in Figure 9d (magnification of the chosen part of spectral characteristics) which is not observed for 1550* mixture.

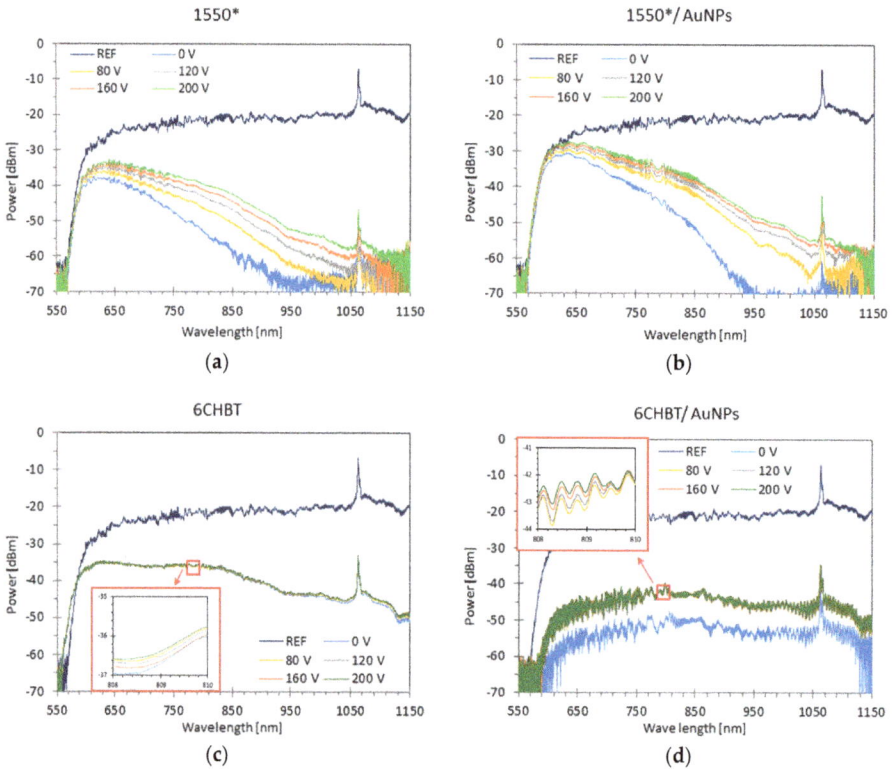

Figure 9. Spectra obtained for LCs: (**a**) 1550* (pure); (**b**) 1550* + 0.1 wt% AuNPs; (**c**) 6CHBT (pure); (**d**) 6CHBT + 0.1 wt% AuNPs, for different steering voltage in the range U of 0–200 V.

If we apply additional modulations of electrical signal—the signal's shape and frequency—we notice that this is mapped on the spectral characteristics. We can select and amplify certain ranges of wavelengths. In practice, this can be used as a bandpass filter. When comparing results for pure LC and with nanoparticles for 1550*, the result is more visible and can be observed for a wider range over 50 nm. Results for a 6CHBT donation of NPs destroy the possibilities of a transfer modulated signal (see Figure 10).

The application of different LC mixtures influences speed of device operation and answer time to measurement parameters). The scheme of the optical system for measuring time courses depending on voltage and temperature is shown in Figure 11. For the measurement we used as a source of light a single mode fiber pigtailed laser (LP852-SF30, ThorLabs, Newton; USA) operating at the center wavelength of 846 nm, a signal generator (DG1022, RIGOL, Beijing; China), voltage amplifier (A400D, FLC Electronics, Molndal; Sweden), an Si-amplified detector operating in the wavelength range of 350–1100 nm, and a climatic chamber (VCL 7010, Votsch Industrietechnik, Balingen, Germany) for temperature measurement in the temperature range of −70 to 180 °C.

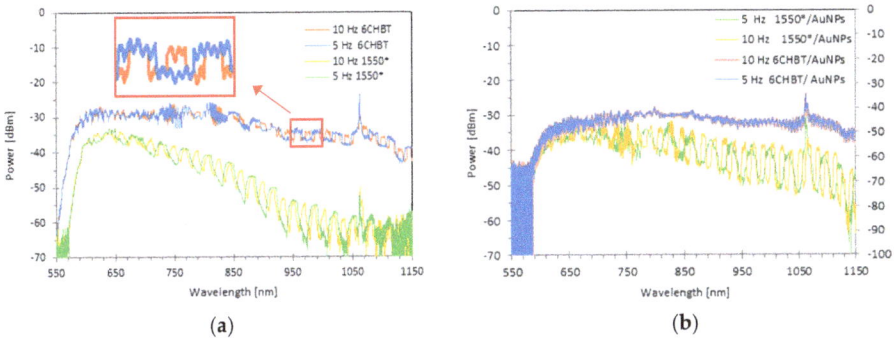

Figure 10. Spectral characteristics in the wavelength range of 550–1150 nm for applied 200 V signal amplitude. The square shape of modulation with a frequency of 5 Hz and 10 Hz has been applied (**a**) pure liquid crystals; (**b**) liquid crystals doped with 0.1 wt% AuNPs.

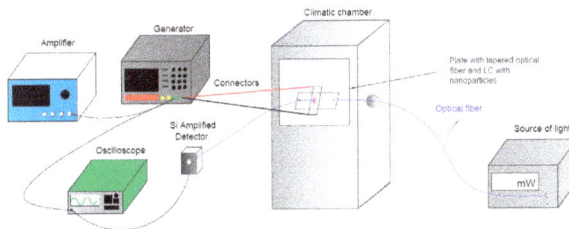

Figure 11. Scheme of the optical system for measuring time courses depending on voltage and temperature.

In a taper device with LC, answer time can be translated into the switching on and off times of the liquid crystal material (times of molecules director reorientation). Time courses obtained for a steering voltage range of U = 80–200 V at room temperature (25 °C) are shown in Figure 12. As can be seen, doping NPs to 1550* mixture made a better image of square modulation with a higher power level. For all voltage measurement, switching times on and off can be easy to measure as shown in Figure 12a,b. For 6CHBT, doping NPs made increasing losses, the signal stays flat for pure and doped LC.

Investigation on LC devices was also provided for temperature dependence (see Figure 13). For both LC 1550* and 6CHBT we measured temperature in the range of 25–40 °C with a square modulation of 10 Hz and a voltage steering of 200 V which determined full molecules' reorientation—above this voltage, there was no increase of power level.

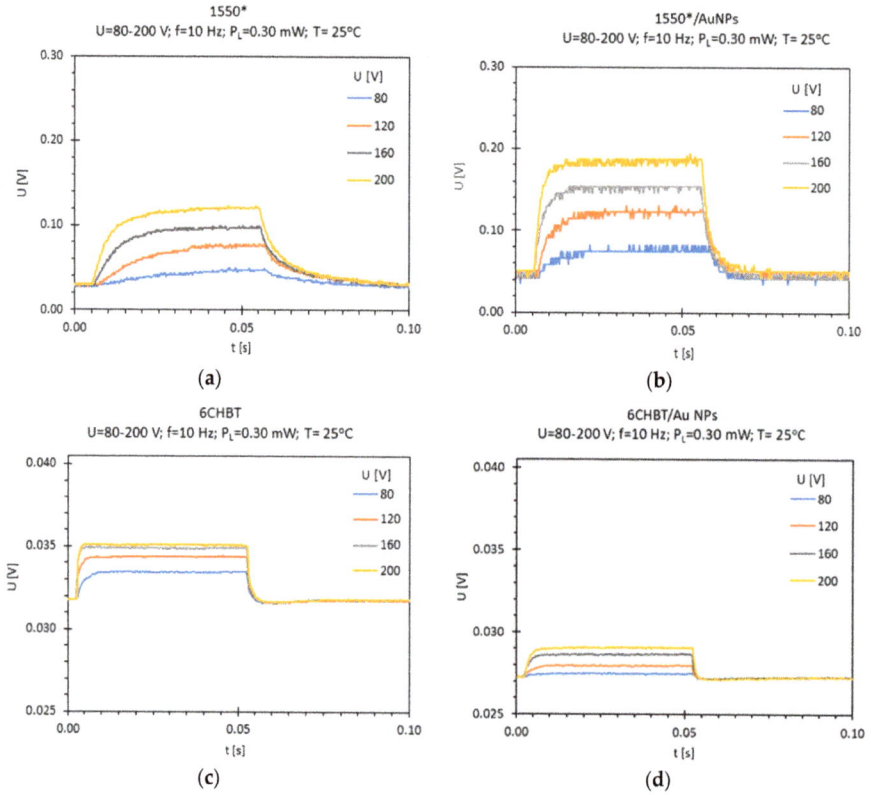

Figure 12. Time courses obtained for the steering voltage range of U = 80–200 V at room temperature (25 °C) for: (**a**) 1550* (pure); (**b**) 1550* + 0.1 wt% AuNPs; (**c**) 6CHBT (pure); (**d**) 6CHBT + 0.1 wt% AuNPs.

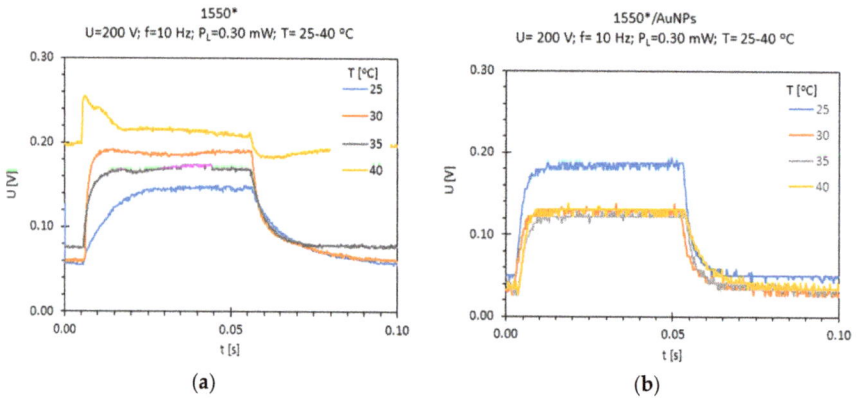

Figure 13. *Cont.*

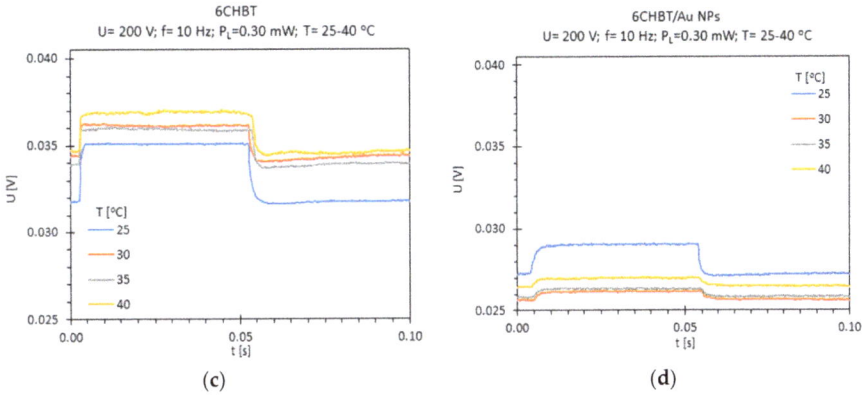

Figure 13. Time courses obtained for the steering voltage U = 200 V at the temperature range of T = 25–40 °C for: (**a**) 1550* (pure); (**b**) 1550* + 0.1 wt% AuNPs; (**c**) 6CHBT (pure); (**d**) 6CHBT + 0.1 wt% AuNPs.

As can be seen in Figure 13, for 1550* LC together with a temperature power increase of transmitted light also increases. For higher temperature modulation, the signal mapped by the LC taper device does not follow the forcing signal. Such changes can be a result of the reduction of elastic and viscosity parameters between the molecules of the LC together with temperature increases, as well as a change of the effective value of the refractive index (medium molecules director). For the 1550* LC with NPs' donation, we observed a step decrease of power for 30 °C, and above this temperature, there is no change of power level which can be strictly connected with the NPs and their influence on LC structure and molecules' direction. For pure and doped 6CHBT LC, we observed the inversion of a power level change. For pure LC together with increasing temperature, power increases, and for the doped one with temperature increase power decreases.

As a switch on (off) time we assume the time required for a change in the light transmission through the investigated liquid crystal cell from 10% to 90% (or vice versa) of its maximal value when steering voltage is switched on (or off) (See Figure 14a,b).

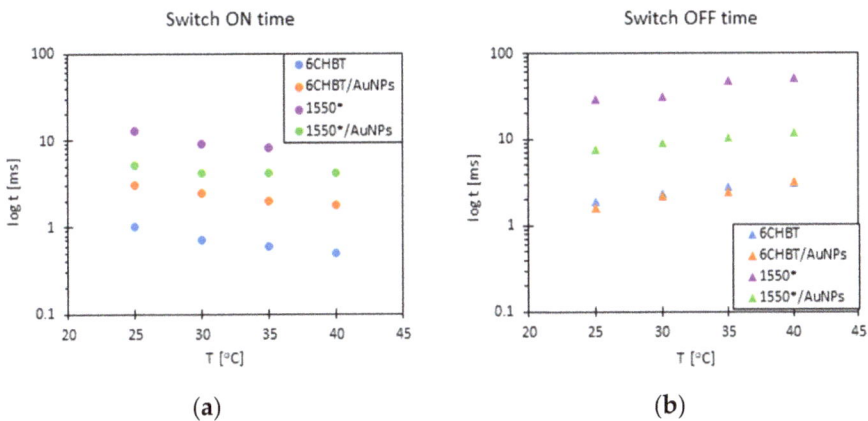

Figure 14. (**a**) Switch ON and OFF time for a for 1550* and (**b**) 6CHBT (pure), 1550* + 0.1 wt% AuNPs; and 6CHBT + 0.1 wt% AuNPs.

For both cases, the applied pure 1550* mixture possesses the highest switch on (Figure 14a) and switch off times (Figure 14b). By doping this mixture with Au nanoparticles, we obtain time decrease several times. For 6CHBT we can observe that switch on time for the doped LC increases over three times in contrast to the pure LC. Switching off times are almost on the same level in this liquid crystal.

Comparing the times of switch on and off of LC 1550* and 6CHBT with nanoparticles, it can be observed that 6CHBT is much faster than the 1550* LC mixture which is connected with the viscosity strength of the chosen LC.

4. Conclusions

In this article, we present the possibility of manufacturing optical devices using two technologies: taper optical fibers and liquid crystals with gold nanoparticles. The optical fiber taper process allows for direct influence on propagating light beam inside the structure by changing the boundary condition of cladding parameters. This technology connected with LC materials, enables us to manufacture in-line miniaturized devices sensitive for many factors like temperature, electric, and magnetic field. Liquid crystals as materials possessing anisotropic properties are sensitive to changes of external factors: temperature, electric and magnetic fields. Any change in the external factor is associated with a change of the liquid crystal material properties. The light in the optical fiber taper enables a detection of the mentioned changes due to their sensitivity to a changing refractive index of the surrounding material. When we use only the electric field, we operate on the coefficient change in the n_e–n_o range at constant temperature. When the temperature is changed, the values ne and no are simultaneously reduced. After exceeding the transition temperature to the isotropic state, n_{iso} is below the n of glass. Taking into consideration the conducted state of the art tests, we notice that as temperature rises, the spectral characteristics change as a result of structural change and such parameters of the liquid crystal as viscosity, refractive index, elastic. It is difficult to separate the result of the measured parameters in optical fiber measurements; however, in our case, it can be possible to separate two factors since we know the answer to a given temperature and voltage in approximation. Depending on the possible application, the working parameters for signal detection with different times of on/off switching can be improved. Connection of liquid crystals with nanoparticles allows for a significant change of their optical parameters, including effective refractive index (the medium molecule's director) and the materials' parameters like viscosity. LC is one of the materials which can be used as a functional material enabled to detect many factors like the electric or magnetic field or temperature. The results with LC and nanoparticles described show the possibility of manufacturing in line miniaturized devices for commercial application. By applying different kinds of LC mixtures, we can adjust the parameters of the devices to optimize their work depending of measurement requirements.

Author Contributions: Conceptualization L.R.J. and K.A.S.; methodology K.A.S.; J.E.M.; J.K.; B.J. and B.B., formal analysis J.E.M. and K.A.S., investigation J.E.M.; J.K. and K.A.S., data curation K.A.S. and L.R.J, writing—original draft preparation J.E.M.; J.K. and K.A.S., writing—review and editing L.R.J., visualization J.E.M. and J.K., supervision L.R.J. and K.A.S., funding acquisition J.E.M. and L.R.J.

Funding: This research was funded by National Science Centre, grant number UMO-2018/29/N/ST7/02347, Military University of Technology grant number RMN 08/689 as well as Statutory Task PBS 898 and supported by the Program of the Republic of Poland—Research Grant MUT project no. GBMON/13-995/2018/WAT in 2019.

Conflicts of Interest: The authors declare no conflict of interest.

References

1. Nealon, G.L.; Greget, R.; Dominguez, C.; Nagy, Z.T.; Guillon, D.; Gallaniand, J.L.; Donnio, B. Liquid-crystalline nanoparticles: Hybrid designand mesophase structures. *Beilstein J. Org. Chem.* **2012**, *8*, 349–370. [CrossRef] [PubMed]

2. Škarabot, M.; Ryzhkova, A.V.; Muševič, I. Interactions of single nanoparticles in nematic liquid crystal. *J. Mol. Liq.* **2018**, *267*, 384–389. [CrossRef]

3. Hsu, C.J.; Lin, L.J.; Huang, M.K.; Huang, C.Y. Electro-optical Effect of Gold Nanoparticle Dispersed in Nematic Liquid Crystals. *Crystals* **2017**, *7*, 287. [CrossRef]
4. Woliński, T.R.; Siarkowska, A.; Budaszewskia, D.; Chychłowskia, M.; Czapla, A.; Ertman, S.; Lesiak, P.; Rutkowska, K.; Orzechowski, K.; Sala-Tefelska, M.; et al. Recent advances in liquid-crystal fiber-optics and photonics. *Proc. SPIE* **2017**, *10125*, 101250W. [CrossRef]
5. Kowerdziej, R.; Garbat, K.; Walczakowski, M. Nematic liquid crystal mixtures dedicated to thermally tunable terahertz devices. *Liq. Cryst.* **2018**, *45*, 1040–1046. [CrossRef]
6. Sengupta, A. Liquid Crystal Theory. In *Topological Microfluidics: Nematic Liquid Crystals and Nematic Colloids in Microfluidic Enviroment*; Springer: Basel, Switzeland, 2013; pp. 7–34.
7. Katsunari, O. Wave Theory of Optical Waveguides. In *Fundamentals of Optical Waveguides*; Academic Press: Londyn, UK, 2006; pp. 1–12.
8. Frazão, O.; Santos, J.L.; Araújo, F.M.; Ferreira, L.A. Optical sensing with photonic crystal fibers. *Laser Photonics Rev.* **2008**, *2*, 449–459. [CrossRef]
9. Brambilla, G. Optical fibre nanowires and mikrowires: Review. *J. Opt.* **2010**, *12*, 043001. [CrossRef]
10. Tian, Y.; Wang, W.; Wu, N.; Zou, X.; Wang, X. Tapered Optical Fiber Sensor for Label-Free Detection of Biomolecules. *Sensors* **2011**, *11*, 3780–3790. [CrossRef]
11. Moś, J.E.; Stasiewicz, K.A.; Garbat, K.; Morawiak, P.; Piecek, W.; Jaroszewicz, L.R. Tapered fibre liquid crystal tunable broad band filter. *Phys. Scr.* **2018**, *93*, 125002. [CrossRef]
12. Korec, J.; Stasiewicz, K.A.; Strzeżysz, O.; Kula, P.; Jaroszewicz, L.R. Electro-Steering Tapered Fiber-Optic Device with Liquid Crystal Cladding. *J. Sens.* **2019**, *2019*, 1617685. [CrossRef]
13. Chao, D.; Wang, Q.; Zhao, Y. Electrically tunable long period gratings temperature sensor based on liquid crystal infiltrated photonic crystal fibers. *Sens. Actuators A Phys.* **2018**, *278*, 78–84.
14. Czapla, A.; Bock, W.J.; Woliński, T.R.; Mikulic, P.; Dąbrowski, R.; Nowinowski-Kruszelnicki, E. Electically tunable long-period fiber gratings with low- birefringence liquid crystal near the turn-around point. *Opto-Electron. Rev.* **2017**, *25*, 290–295. [CrossRef]
15. Choudhury, P.K.; Soon, W.K. On the tapered optical fibers with radially anisotropic liquid crystal clad. *Prog. Electromagn. Res.* **2011**, *115*, 461–475. [CrossRef]
16. Wahle, M.; Kitzerow, H.S. Liquid crystal assisted optical fibers. *Opt. Express* **2014**, *22*, 262–273. [CrossRef] [PubMed]
17. Woliński, T.R.; Szaniawska, K.; Ertman, S.; Lesiak, P.; Domański, A.W.; Dąbrowski, R.; Nowinowski-Kruszelnicki, E.; Wojcik, J. Influence of temperature and electrical fields on propagation properties of photonic liquid-crystal fibers. *Meas. Sci. Technol.* **2006**, *17*, 985–991. [CrossRef]
18. Larsen, T.T.; Bjarklev, A. Optical devices based on liquid crystal photonic bandgap fibres. *Opt. Express* **2003**, *11*, 2589–2596. [CrossRef]
19. Veilleux, C.; Lapierre, J.; Jacques, B. Liquid crystal clad tapered fibers. *Opt. Lett.* **1986**, *11*, 733–735. [CrossRef]
20. Veilleux, C.; Black, R.J.; Lapierre, J. Nematic liquid crystal clad tapered optical fiber with temperature sensing properties. *J. Appl. Phys.* **1990**, *67*, 6648–6653. [CrossRef]
21. Dąbrowski, R.; Garbat, K.; Urban, S.; Woliński, T.R.; Dziaduszek, J.; Ogrodnik, T.; Siarkowska, A. Low-birefringence liquid crystal mixtures for photonic liquid crystal fibres application. *Liq. Cryst.* **2017**, *44*, 1911–1928. [CrossRef]
22. Zhang, L.; Lou, J.; Tong, L. Micro/Nanofiber Optical Sensors. *Photonic Sens.* **2011**, *1*, 31–42. [CrossRef]
23. Polynkin, P.; Polynkin, A.; Peyghambarian, N.; Mansuripur, M. Evanescent field-based optical fiber sensing device for measuring the refractive index of liquids in microfluidic channels. *Opt. Lett.* **2005**, *30*, 1273–1275. [CrossRef] [PubMed]
24. Choudhury, P.K.; Soon, W.K. On the transmission by liquid crystal tapered optical fibers. *Optik* **2011**, *122*, 1061–1068. [CrossRef]
25. Laudyn, U.; Rutkowska, K.; Rutkowski, R.T.; Karpierz, M.A.; Woliński, T.R.; Wójcik, J. Nonlinear effects in photonic crystal fibers filled with nematic liquid crystals. *Cent. Eur. J. Phys.* **2008**, *6*, 612–618. [CrossRef]
26. Jadżyn, J.; Hellemans, L.; Czechowski, G.; Legrand, C.; Douali, R. Dielectric and viscous properties of 6CHBT in the isotropic and nematic phases. *Liq. Cryst.* **2010**, *27*, 613–619. [CrossRef]
27. Siarkowska, A.; Chychłowski, M.; Budaszewski, D.; Jankiewicz, B.; Bartosewicz, B.; Woliński, T.R. Thermo- and electro-optical properties of photonic liquid crystal fibers doped with gold nanoparticles. *Beilstein J. Nanotechnol.* **2017**, *8*, 2790–2801. [CrossRef] [PubMed]

28. Budaszewski, D.; Chychłowski, M.; Budaszewska, A.; Bartosewicz, B.; Jankiewicz, B.; Woliński, T.R. Enhanced efficiency of electric field tunability in photonic liquid crystal fibers doped with gold nanoparticles. *Opt. Express* **2019**, *27*, 14260–14269. [CrossRef] [PubMed]

29. Brust, M.; Walker, M.; Bethell, D.; Schiffrin, D.J.; Whyman, R. Synthesis of thiol-derivatised gold nanoparticles in a two-phase liquid-liquid system. *J. Chem. Soc. Chem. Commun.* **1994**, 801–802. [CrossRef]

30. Stasiewicz, K.A.; Jaroszewicz, L.R. Automatic set-up for advanced optical fiber elements manufacturing. *Proc. SPIE* **2005**, *5952*. [CrossRef]

31. Weglowski, R.; Piecek, W.; Kozanecka-Szmigiel, A.; Konieczkowska, J.; Schab-Balcerzak, E. Poly(esterimide) bearing azobenzene units as photoaligning layer for liquid crystals. *Opt. Mater.* **2015**, *49*, 224–229. [CrossRef]

32. Orlandi, S.; Benini, E.; Miglioli, I.; Evans, D.R.; Reshetnyak, V.; Zannoni, C. Doping liquid crystals with nanoparticles. A computer simulation of effects of nanoparticle shape. *Phys. Chem.* **2016**, *18*, 2428–2441. [CrossRef]

33. Urbanski, M. On the impact of nanoparticle doping on the electro-optic response of nematic hosts. *Liq. Cryst. Today* **2015**, *24*, 102–115. [CrossRef]

Article

Multifrequency Driven Nematics

**Noureddine Bennis *, Jakub Herman, Aleksandra Kalbarczyk, Przemysław Kula
and Leszek R. Jaroszewicz**

Faculty of Advanced Technologies and Chemistry, Military University of Technology, 2 gen. S. Kaliskiego St.,
00-908 Warsaw, Poland; jakub.herman@wat.edu.pl (J.H.); aleksandra.kalbarczyk@wat.edu.pl (A.K.);
przemyslaw.kula@wat.edu.pl (P.K.); leszek.jaroszewicz@wat.edu.pl (L.R.J.)
* Correspondence: noureddine.bennis@wat.edu.pl

Received: 10 May 2019; Accepted: 23 May 2019; Published: 27 May 2019

Abstract: Liquid crystals act on the amplitude and the phase of a wave front under applied electric fields. Ordinary LCs are known as field induced birefringence, thus both phase and amplitude modulation strongly depend on the voltage controllable molecular tilt. In this work we present electrooptical properties of novel liquid crystal (LC) mixture with frequency tunable capabilities from 100Hz to 10 KHz at constant applied voltage. The frequency tunability of presented mixtures shown here came from composition of three different families of rodlike liquid crystals. Dielectric measurements are reported for the compounds constituting frequency-controlled birefringence liquid crystal. Characterization protocols allowing the optimum classification of different components of this mixture, paying attention to all relevant parameters such as anisotropic polarizability, dielectric anisotropy, and dipole moment are presented.

Keywords: nematic; dual frequency nematic; dielectric anisotropy; optical modulation

---

1. Introduction

Spatial light modulators (SLMs) based on liquid crystals (LCs) require very specific combinations of physical properties of liquid crystal mixtures to operate efficiently, requiring different combinations of components. Multicomponents mixtures of LCs find widespread use in photonic applications. The reason for this, is the wide temperature range necessary for applications [1]. The more important physical properties of LCs mixtures include electric permittivities, optical refractive indices, elastic constants, and viscosities. In order to understand the dielectric properties of LCs mixtures, some knowledge of the behavior of each component of the mixture is necessary. Owing to their electrical and optical properties, they are very sensitive to external electric and magnetic fields and they play an important role in many types of photonic devices enabling the development of multiphase spatial light modulators (SLMs) that can perform high-resolution, dynamic, optical beam positioning as well as temporal and spatial beam shaping. The optical system of such as spatial light modulators can perform with extremely high-resolution having a very small pixel size of about a few μm. This small resolution makes the latest generation of microdisplays based on LCs useful for many applications. However, we shall be concerned with negative physical effects related to smaller pixel sizes because of the inherent elasticity of the liquid crystal (LC) material, the surface anchoring of the alignment layer, and the fringing field of the discrete electrodes' voltage distribution [2]. The electrical field that exists between a pixel turned on and its neighbor at a different voltage causes part of the liquid crystal molecules to adopt a tilt angle opposite to the one in the main part of the pixel [3]. In order to avoid the fringing field effect originating from different values of voltage distribution, a digital driving frequency such as a pulse width modulation is needed. In this case, equal voltage with a predetermined frequency should be applied to adjacent pixel electrodes. As well as the amplitude of applied voltage could alter the reorientation of LC molecules, also its frequency may alter their response. Thus, we need to understand the most important information about the molecular dynamics of LCs in the presence of the alternating electric field. When a highly anisotropic molecules of LCs are found in an electric field their electric polarization depends on their orientation with respect to it. The frequency dependence can be described from the physical origin of the induced polarization of the molecules. This induced polarization interacts with an external electric field and the torque tries to align the long molecular axis parallel or perpendicular to the external electric field depending of the sign of the dielectric anisotropy. If the applied field varies with time, then the frequency dependence of the permittivity is an additional property of the LCs material under study. Thus at much higher frequency of the applied electric field, the molecular dipoles do not rotate fast enough to contribute to the dielectric response, therefore the phase shift between the electric field E and electric polarization occurs. This delay yields to the dielectric relaxation of the LC molecules, causing the dielectric dispersion and losses [4]. For the uniaxial nematic LCs there are two independent components of the electric permittivity, parallel $\varepsilon_{\parallel}$ and perpendicular $\varepsilon_{\perp}$ to the axis of symmetry. Both principal components of the dielectric permittivity tensor show different frequency dependencies and temperature dependencies as well [5–7]. Perpendicular component $\varepsilon_{\perp}$ exhibits the high frequency process which can be treated as an ordinary Debye dispersion of fluid systems which usually occurs in the microwave region. However, $\varepsilon_{\parallel}$ may exhibit in some cases an additional dispersion at much lower frequencies in the radio-frequency region. This low-frequency dispersion of $\varepsilon_{\parallel}$ is caused by orientational relaxation of the permanent dipole moment of the parallel component to the long molecular axis. Maier and Meier are the first who formulated these ideas more qualitatively by applying Onsager's theory of static polarization to nematic LCs [8]. Measurements of the dielectric properties of LCs give relaxation frequencies, which reflect dynamic processes involving a change in electric polarization that results from the chemical structure of the mesogen forming the LC material. Many mixtures of LCs exhibit a single dielectric relaxation even if the components individually have relaxations in different frequency ranges. However, in some mixtures of mesogens having significantly different structures, for example mixtures of two-ring and three-ring mesogens, two separated relaxations are sometimes detected [9]. The possibility of designing mixtures with very low relaxation frequencies is useful for a dual frequency addressing displays [10]. Such mixture is usually formed by a combination of many components such as molecules having large transverse dipole moment with $\Delta\varepsilon < 0$ and molecules with a large longitudinal dipole moment with $\Delta\varepsilon > 0$ [11]. The availability of a dielectric anisotropy of either sign depending on the frequency of the applied field allows the possibility to improve the dynamic response of electro-optical devices. One of the main drawbacks of Dual frequency liquid crystal (DFLC) was the high crossover frequencies; problems arising when high frequencies are applied (> 50 kHz) are higher cost in the driving circuit, larger heat dissipation, and stronger dielectric heating [12]. For this reason, the research on materials with low crossover frequencies led to novel materials [13–15]. In this work, we report on LCs mixture design by mixing three different families of LC compounds. Two groups of the components are dielectrically positive nematics, having different generated dipole moment, and one group of compounds is dielectrically neutral, which has been chosen as viscosity adjusting component. The resulting mixture meets the requirements of the large positive dielectric anisotropy LCs to be continuously controlled by low frequencies of the electrical applied voltage. The dielectric anisotropy of the proposed mixture goes to zero instead of being negative at high frequencies of the electric field. The mesomorphic properties of the three components of the mixture under study and electro-optical features in new formulated mixture are reported.

2. Influence of NLC Mixtures on Dielectric Relaxation

Particular semiempirical models have been regularly used in the literature to allow the prediction of the electrooptical parameters such as the dielectric anisotropy $\Delta\varepsilon$ and the birefringence Δn of dielectric materials. These two parameters can be derived from single molecule parameters, such as the dipole moment μ and the polarizability α, using the Maier–Meier theory [8]. According to this theory, the supramolecular parameter $\Delta\varepsilon$ can be connected to the principal component of the molecular polarizability and the value and angular position of permanent electric dipole moment μ.

Dielectric relaxation in LC material is influenced by the molecular structure used in the liquid crystal mixture and also by other factors such as intermolecular interactions and the local viscosity and temperature. This dependence gives raise to the possibility of designing liquid crystal mixtures with very low relaxation frequencies of electric permittivity parallel $\varepsilon_{\parallel}$, which is useful for many applications [16]. Apart from traditional SLM, the unique properties of this LC mixture can be used in all kind of optical phase modulators recently reported, e.g., adaptive lenses [17], beam steering [18], correction of aberrations [19], 3D vision applications [19–22], novel aberrations correctors for rectangular apertures [23], microaxicon arrays [24], multioptical elements [25], hole-patterned microlenses [26], multifocal microlenses [27], high fill-factor microlenses [28], frequency controlled [29] microlenses, optical vortices [30], lensacons, and logarithmic axicons [31]. The LC mixtures used in this work have been designed by mixing three different families of LC compounds to meet the requirements of the LC to be controlled by low frequencies of the electrical applied voltage. We have formulated the mixture using the set of compounds belonging to three different groups (see Table 1).

Table 1. General molecular structures of compounds used to form the investigated nematic mixture and their weight %.

Components	Chemical Structure	Weight %
I		6.5
II		63.5
III		30

The investigated material is a nematic composition of three different families of rod like LCs. The chemical formulas and the bond dipole moments of different groups forming the multifrequency driven nematic are shown in Figure 1. These structures belong to fluorine substituted 4-[(4-cyanophenoxy)carbonyl]phenyl 4-alkylbenzoates [32] (component I), fluorine substituted alkyl-alkyl phenyl-tolanes [32–34] (component II), and alkyl-alkyl bistolanes [35,36] (component III). Each component is characterized by their strong nematogenic character.

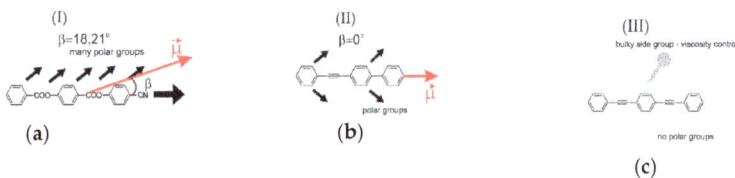

Figure 1. The structures of the three different groups forming the mixture: (**a**) component I, (**b**) component II and (**c**) component III. Red arrows indicate the position of the molecular dipole moments.

Two of the components are dielectrically positive nematics and one is dielectrically neutral. Components I and II have a large longitudinal dipole moment. Group I structures (Figure 1a) show a strong dipole moment that can be estimated as a vector sum of the single moments of all intramolecular chemical bonds. Their positions and orientations within the molecule for these dipole moments are additives, having substantial influence on nematic ordering. The largest contribution to the dipole moment of component I comes from the cyano group. This in turn is coupled with two ester groups that additionally enhance dipole moment value. The longitudinal electric dipole moment of the component (I) is about 12,4 debye due to a triple −CN bond; the permanent dipole moment forms an angle with the molecular axis equal to 18.21°. This group of compounds is characterized by dielectric anisotropy $\Delta\varepsilon$ in the range from 60 to 80, which can be realized by the use of polar chemical groups, making this component a highly polar nematics liquid crystal. Additionally component I shows moderate molecular anisotropy and moderate electronic polarizability [36]. These highly polar LCs with high positive dielectric anisotropies tend to operate at lower voltages. Medium birefringent component II used in this mixture has a dipole moment of about 3 debye and $\Delta\varepsilon$ is in the range from +2 to +4 at 1 kHz. It has low melting enthalpies in order to enhance an excellent solubility with other two groups of components of the mixture. The group III is mainly used as diluters with which one may maintain the proper viscosity of LCs mixtures [32–34]. This component does not consist of polar substituents nor polar groups, therefore is a nonpolar highly birefringent liquid crystal. Bistolanes III show high anisotropy of molecular polarizability, as a result of a strongly conjugated system of benzene rings and carbon-carbon triple bonds [37]. In this component there is only an induced polarization that consists of two parts: the electronic polarization and the ionic polarization. Its dipole moment is μ=0. Thus, its $\Delta\varepsilon$ is expected to be small ($\Delta\varepsilon$ = +0,7 at 1 kHz) and is considered as medium to high birefringent with high nematogenity. This component is laterally substituted with short alkyl chains (methyl, ethyl, trifluoromethyl), shows strong nematic character, and therefore it minimizes the smectic formation in multicomponent systems under study. In contrast, in the LCs with polar molecules groups (I) and (II), there is, in addition to the total induced polarization, the orientation polarization due to the tendency of the permanent dipole moments to orient themselves parallel to the electric field.

3. Dielectric Measurements Results

In the context of the dielectric properties of the three components forming the multifrequency driven LCs mixture, their dielectric properties are of interest from a number of aspects. In this section we will focus on their dielectric relaxation spectra. the characterization involves the measure of the electric permittivity tensor over a range of frequencies from 100 Hz to 1 MHz. Dielectric investigation was carried out on the three components forming the mixture in planar aligned cells with thicknesses of 0.7 mm. The alignment was induced on the substrates by spin-coating polyimide SE-130 (Nissan Chemical Industries, Ltd., Tokyo, state, Japan), then baking at 180 °C for 1 h. The coated substrates were rubbed in the same direction and were assembled in antiparallel orientation. For this measurement, an external magnetic field was applied to the cell in order to orient the LCs molecules. In this case, two experimental geometries were applied; the parallel and perpendicular orientations of the director with respect to the electric field have been achieved by turning the electromagnet by 90° enabling the measurement of the $\varepsilon_{\parallel}$ and $\varepsilon_{\perp}$ which are parallel and perpendicular permittivity tensor components, respectively. The frequency dependences of the real (upper graphs) and imaginary (lower graphs) parts of the electric permittivity measured for the nematic phase of the three groups are shown in Figure 2.

The component (I) has relatively long molecules, therefore the resulting mesophases tend to be of high viscosity and are difficult to align; it has high melting and clearing temperatures and accordingly has very high viscosity which makes difficult the characterization of their dielectric properties. Therefore, the dielectric properties of this compound have to be measured in other neutral nematic matrices in order to extrapolate their resulting dielectric properties. In Figure 2a, only 10% molar concentration of component (I) has been used. The dielectric results of the pure material would have about 10 times higher $\Delta\varepsilon$, 10 times higher $\varepsilon_{\parallel}$, and more than 10 times lower relaxation frequency. For this component, the relaxation processes are detected at low frequencies and are more characteristic of collective modes normally associated with orientational relaxation of the permanent dipole moments. Components (II) and (III) have already much lower crystalline nematic transition temperatures, and therefore, they could be evaluated as pure systems. Those materials were tested at 51 °C and 58 °C respectively. Under given temperatures, these components are far from nematic-isotropic transition, in consequence their properties within the nematic phase are flattened. The dielectric permittivity constants $\varepsilon_\parallel$ and $\varepsilon_\perp$ of the component III do not depend on frequency, therefore, dielectric losses are very small in the frequency range [100 Hz,15 KHz]. Figure 2 shows that the constituents of the formulated mixture have different dielectric responses to an external electric field of variable frequency. Note that the frequency dependence of the parallel component of electric permittivity is different for each group. We have formulated the mixture using the set of compounds belonging to three different families (See Table 2).

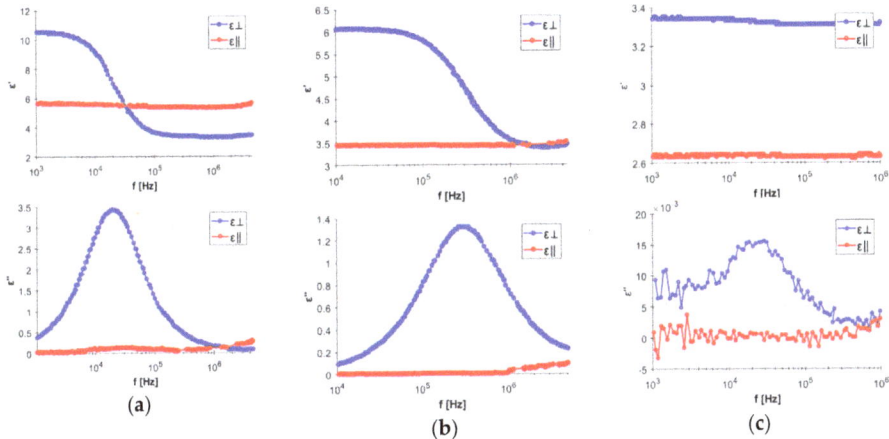

Figure 2. Frequency dependence of the real part (upper) and imaginary part (lower) of the complex permittivity of the formulated mixture in the measuring frequency range 100Hz to 1MHz for the three different components of the mixture: (**a**) component (I), (**b**) component (II) and (**c**) component (III).

Table 2. Mesomorphic properties of the three components of the mixture under study.

Group	Polarity	Anisotropic Polarizability	Dipole Moment	Dielectric Anisotropy @ 1KHz	Viscosity
I	high	moderate	12,4D	(60–80)	very high
II	moderate	high	3D	(2–4)	low
III	very low	high	0D	0,7	moderate

Figure 3 shows the frequency dependence of the real and imaginary part of the complex permittivity of the formulated mixture in the measuring frequency range 100Hz to 1MHz. These results show that the mixing up of the three groups results in a positive nematic mixture with strong dependence on the frequency of the applied electric field. At a low frequency regime, the nematic mixture exhibits a large positive dielectric permittivity of about 9.5 and above 100Hz, $\varepsilon_\parallel$ starts to decrease linearly as frequency increases. However, $\varepsilon_\perp$ is kept constant at a value of 3.5. In the frequency range 100Hz to 1MHz, both $\varepsilon_\parallel$ and $\varepsilon_\perp$ stay at about 3,5. As a result, a continuously varying dielectric anisotropy, from 6 to 0 in the frequency range 100Hz-15KHz, is obtained. This mixture could be employed in optical devices which can readily be used as multiphase spatial light modulators controlled by a frequency.

The dielectric spectra showed in Figure 3a can be presented in the form of a Cole–Cole plot. This diagram represents a series of two interlinked semicircles whose radius of the semicircle are inversely proportional to the relaxation time, indicating two relaxation mechanisms separated in frequency. These results may then be explained as different dynamic modes associated with relaxation of different dipoles group, each relaxing at a different frequency. Assuming that, these contributions to the electric permittivity occur at sufficiently different frequencies, and therefore, they can be separated in the dielectric spectrum. The dashed lines in Figure 3c mark two Debye-type processes fitted to the spectra. From those results, it seems that for our mixture two relaxation processes appear in the Cole–Cole plot corresponding to frequencies 3 KHz and 34 KHz, which show evidence of two or more separate relaxation processes.

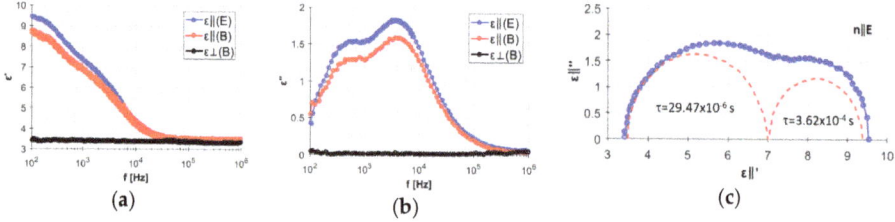

Figure 3. Frequency dependence of (**a**) real and (**b**) imaginary parts of the complex permittivity plotted in the measuring frequency range 100 Hz to 1MHz. (**c**) Cole–Cole plot showing relaxation process with two time constants.

Figure 4 shows the results of the optical retardance versus the alternative voltage waveform applied. For optical retardance measurements, we have used cells prepared with the same alignment method described above. However, the thickness of the assembled cells has been reduced 5 μm and next filled by the formulated mixture (Figure 4a) and by conventional liquid crystal 6CHBT (Figure 4b) when the voltage varies from 0 V to 15 V at a constant frequency. A maximal phase modulation depth obtained at λ = 6323 nm is 5π in case of 5005 and $2.7\,\pi$ in case of 6CHBT. Upon changing the frequency of the applied signal, a change in the retardance modulation is observed in the formulated mixture. This change is determined by the tunability of dielectric anisotropy by the frequency of the applied electric field. However, for LCs 6CHBT, since the relaxation frequency occurs in the 100 MHz region [7] no dependence of optical retardation on the frequency has been detected.

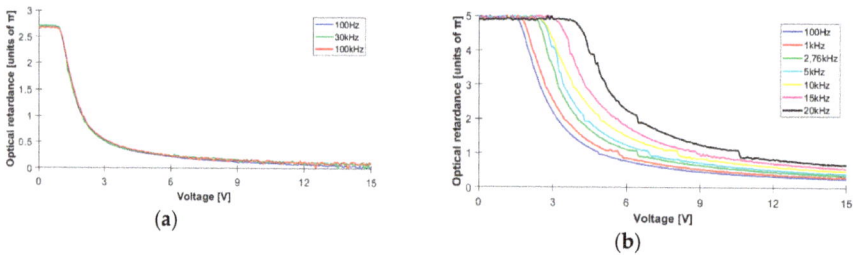

Figure 4. Voltage-dependent phase changes at different frequencies measured in 5 μm cell filled with: (**a**) formulated mixture, (**b**) 6CHBT liquid crystals.

These results show that as the frequency increases the electro-optical characteristics of the cell filled with multifrequency driven nematics shifts toward the higher operation voltage. The frequency dependence of the threshold voltage shown in Figure 4a reflects the frequency dependence of the dielectric anisotropy of the proposed LCs material. This frequency effect mainly originates from frequency dependence of dielectric permittivity $\Delta\varepsilon$ since all other parameters are independent of frequency.

4. Conclusions

Dielectric and electrooptical measurements were carried out on the novel liquid crystal mixture with frequency tunable capabilities from 100 Hz to 10 KHz. The dielectric anisotropy of the proposed mixture goes to zero at higher frequencies instead of being negative, which is typical for dual frequency nematics (DFN). Characterization protocols allowing the optimum classification of different components of this mixture, paying attention to all relevant parameters such as anisotropic polarizability and dielectric anisotropy have been presented. The dielectric permittivity of the mixtures based on the three components depends upon the polarity of each particular group and their concentrations in the mixture. These studies can aid the development of new liquid crystals materials with carefully tailored dielectric properties.

Author Contributions: Methodology, writing and editing, N.B.; optical retardance measurement, A.K.; synthesis of the liquid crystalline single compounds and formulation of the mixture, J.H.; discussion of the results, P.K.; and supervision of the work, L.R.J.

Funding: Ministry of National Defense of Poland under grant number (GBMON/13-995/2018/WAT), Military University of Technology grant no 23-895.

Acknowledgments: The authors would like to thank Stanisław Urban for the dielectric measurements and his helpful discussions.

Conflicts of Interest: The authors declare no conflict of interest.

References

1. Dąbrowski, R.; Dziaduszek, J.; Garbat, K.; Urban, S.; Filipowicz, M.; Herman, J.; Czerwiński, M.; Harmata, P. Nematic compounds and mixtures with high negative dielectric anisotropy. *Liq Cryst* **2017**, *44*, 1534–1548.
2. Apter, B.; Efron, U.; Bahat-Treidel, A. On the fringing-field effect in liquid-crystal beam-steering devices. *Appl. Opt.* **2004**, *43*, 11–19. [CrossRef] [PubMed]
3. Cuypers, D.; De Smet, H.; Van Calster, A. Electronic compensation for fringe-field effects in VAN LCOS microdisplays. In Proceedings of the International Symposium of the Society-for-Information-Display, Campbell, CA, USA, 18–23 May 2008.
4. Daniel, V. *Dielectric Relaxation*; Academic Press: London, UK; New York, NY, USA, 1967; p. 203.
5. Zakharov, A.V.; Dong, R.Y. Dielectric and elastic properties of liquid crystals. *Phys. Rev. E.* **2001**, *64*, 031701. [CrossRef]
6. Kresse, H. Dielectric Behaviour of Liquid Crystals. In *Advances in Liquid Crystals*; Academic Press: New York, NY, USA, 1983; pp. 109–171.
7. Jadzyn, J.; Hellemans, L.; Czechowski, G.; Legrand, C.; Douali, R. Dielectric and viscous properties of 6CHBT in the isotropic and nematic phases. *Liq Cryst* **2000**, *27*, 613–619. [CrossRef]
8. Maier, W.; Meier, G. Anisotrope dk-dispersion im radiofrequenzgebiet bei homogen geordneten kristallinen flussigkeiten. *Z. Naturforsch* **1961**, *16*, 1200–1205.
9. Toriyama, K.; Sugimori, S.; Moriya, K.; Dunmur, D.A.; Hanson, R. Dielectric study of dipole-dipole interactions in anisotropic solutions. *J. Phys. Chem* **1996**, *100*, 307–315. [CrossRef]
10. Dąbrowski, R.; Celiński, M.; Chojnowska, O.; Kula, P.; Dziaduszek, J.; Urban, U. Compounds with low relaxation frequency and dual frequency mixtures useful for active matrix addressing. *Liq Cryst* **2013**, *40*, 1339–1353.
11. Perkowski, P.; Mrukiewicz, M.; Garbat, K.; Laska, M.; Chodorow, U.; Piecek, W.; Dąbrowski, R.; Parka, J. Precise dielectric spectroscopy of a dual-frequency nematic mixture over a broad temperature range. *Liq Cryst* **2012**, *39*, 1237–1242. [CrossRef]
12. Ibragimov, T.D.; Bayramov, G.M. Application of dual-frequency liquid crystal for tunable selective filtration of infrared radiation. *Optik* **2013**, *124*, 6666–6668. [CrossRef]
13. Song, Q. Fast Response Dual Frequency Liquid Crystal Materials. Ph.D. Thesis, University of Central Florida, Orlando, FL, USA, 2010.
14. Song, Q.; Jiao, M.; Ge, Z.; Xianyu, H.; Gauza, S.; Wu, S.T. High birefringence and low crossover frequency dual frequency liquid crystals. *Mol. Cryst. Liq. Cryst* **2008**, *488*, 179–189. [CrossRef]
15. Xianyu, H.; Gauza, S.; Wu, S.T. Sub-millisecond response phase modulator using a low crossover frequency dual-frequency liquid crystal. *Liq Cryst* **2008**, *35*, 1409–1413. [CrossRef]
16. Kula, P.; Aptacy, A.; Herman, J.; Wojciak, W.; Urban, S. The synthesis and properties of fluoro-substituted analogues of 4-butyl-4′-[(4-butylphenyl)ethynyl]biphenyls. *Liq Cryst* **2013**, *40*, 482–491. [CrossRef]
17. Bailey, J.; Kaur, S.; Morgan, P.B.; Gleeson, H.F.; Clamp, J.H.; Jones, J.C. Design considerations for liquid crystal contact lenses. *J. Phys. D. Appl. Phys.* **2017**, *50*, 485401. [CrossRef]

18. Oton, E.; Morawiak, P.; Mazur, R.; Quintana, X.; Geday, M.A.; Oton, J.M.; Piecek, W. Diffractive and Refractive Liquid Crystal Devices Based on Multilayer Matrices. *J. Light. Technol.* **2019**, *37*, 2086–2093. [CrossRef]

19. Kotova, S.P.; Patlan, V.V.; Samagin, S.A.; Zayakin, O.A. Wavefront formation using modal liquid-crystal correctors. *Phys. Wave Phenom.* **2010**, *18*, 96–104. [CrossRef]

20. Algorri, J.F.; Urruchi del Pozo, V.; Sanchez-Pena, J.M.; Oton, J.M. An Autostereoscopic Device for Mobile Applications Based on a Liquid Crystal Microlens Array and an OLED Display. *J. Disp. Technol.* **2014**, *10*, 713–720. [CrossRef]

21. Algorri, J.F.; Urruchi, V.; Bennis, N.; Sánchez-Pena, J.M. Cylindrical liquid crystal microlens array with rotary axis and tunable capability. *IEEE Electron Device Lett.* **2015**, *36*, 582–584. [CrossRef]

22. Algorri, J.F.; Urruchi, V.; Bennis, N.; Morawiak, P.; Sanchez-Pena, J.M.; Oton, J.M. Integral Imaging Capture System with Tunable Field of View Based on Liquid Crystal Microlenses. *IEEE Photonics Technol. Lett.* **2016**, *28*, 1854–1857. [CrossRef]

23. Algorri, J.F.; Urruchi, V.; Bennis, N.; Sánchez-Pena, J.M.; Otón, J.M. Tunable liquid crystal cylindrical micro-optical array for aberration compensation. *Opt. Express* **2015**, *23*, 13899–13915. [CrossRef] [PubMed]

24. Algorri, J.F.; Urruchi, V.; Bennis, N.; Sánchez-Pena, J.M. Modal liquid crystal microaxicon array. *Opt. Lett.* **2014**, *39*, 3476–3479. [CrossRef]

25. Algorri, J.F.; Love, G.D.; Urruchi, V. Modal liquid crystal array of optical elements. *Opt. Express* **2013**, *21*, 24809–24818. [CrossRef]

26. Urruchi, V.; Algorri, J.F.; Marcos, C.; Sánchez-Pena, J.M. Electrical modeling and characterization of voltage gradient in liquid crystal microlenses. *Rev. Sci. Instrum* **2013**, *84*, 116105. [CrossRef]

27. Algorri, J.F.; Bennis, N.; Urruchi, V.; Morawiak, P.; Sánchez-Pena, J.M.; Jaroszewicz, L.R. Tunable liquid crystal multifocal microlens array. *Sci. Rep.* **2017**, *7*, 17318. [CrossRef]

28. Algorri, J.F.; Urruchi, V.; Bennis, N.; Morawiak, P.; Sánchez-Pena, J.M.; Otón, J.M. Liquid crystal spherical microlens array with high fill factor and optical power. *Opt. Express* **2017**, *25*, 605. [CrossRef]

29. Algorri, J.F.; Bennis, N.; Herman, J.; Kula, P.; Urruchi, V.; Sánchez-Pena, J.M. Low aberration and fast switching microlenses based on a novel liquid crystal mixture. *Opt. Express* **2017**, *25*, 14795. [CrossRef]

30. Algorri, J.F.; Urruchi, V.; Garcia-Camara, B.; Sanchez-Pena, J.M. Generation of Optical Vortices by an Ideal Liquid Crystal Spiral Phase Plate. *IEEE Electron Device Lett* **2014**, *35*, 856–858. [CrossRef]

31. Algorri, J.; Urruchi, V.; García-Cámara, B.; Sánchez-Pena, J. Liquid Crystal Lensacons, Logarithmic and Linear Axicons. *Materials* **2014**, *7*, 2593–2604. [CrossRef]

32. Kula, P.; Dziaduszek, J.; Herman, J.; Dąbrowski, R. Invited Paper: Highly Birefringent Nematic Liquid Crystals and Mixtures. *SID Symposium Digest Technical Papers* **2014**, *45*, 100–103. [CrossRef]

33. Dabrowski, R.; Kula, P.; Herman, J. High Birefringence Liquid Crystals. *Liq Cryst* **2013**, *3*, 443–482. [CrossRef]

34. Weglowska, D.; Kula, P.; Herman, J. High birefringence bistolane liquid crystals: synthesis and properties. *RSC Adv.* **2016**, *6*, 403–408. [CrossRef]

35. Herman, J.; Kula, P. Design of new super-high birefringent isothiocyanato bistolanes-synthesis and properties. *Liq Cryst* **2017**, *44*, 1462–1467. [CrossRef]

36. Ziobro, D.; Kula, P.; Dziaduszek, J.; Filipowicz, M.; Dabrowski, R.; Parka, J.; Czub, J.; Urban, S.; Wu, S.T. Mesomorphic and dielectric properties of esters useful for formulation of nematic mixtures for dual frequency addressing system. *Opto-Elect Rev* **2009**, *17*, 16–19. [CrossRef]

37. Wrobel, S.; Brodzik, M.; Dabrowski, R.; Gestblom, B.; Haase, W.; Hiller, S. Relaxation phenomena in the nematic, nematic reentrant and smectic A phases studied by dielectric spectroscopy. *Mol. Cryst. Liq. Cryst* **1997**, *302*, 223–228. [CrossRef]

crystals

MDPI

Article

Integrated Mach–Zehnder Interferometer Based on Liquid Crystal Evanescent Field Tuning

Manuel Caño-García [1,2,*], David Poudereux [3], Fernando J. Gordo [1], Morten A. Geday [1], José M. Otón [1] and Xabier Quintana [1]

1   Centro de Materiales y Dispositivos Avanzados para Tecnologías de Información y Comunicaciones (CEMDATIC), Universidad Politécnica de Madrid, Av. Complutense 30, 28040 Madrid, Spain; fernando.gordo@upm.es (F.J.G.); morten.geday@upm.es (M.A.G.); jm.oton@upm.es (J.M.O.); x.quintana@upm.es (X.Q.)

2   Department of Nanophotonics, Ultrafast Bio- and Nanophotonics Group, INL-International Iberian Nanotechnology Laboratory, Av. Mestre Jose Veiga s/n, 4715-330 Braga, Portugal

3   Alter Technology TÜV Nord S.A.U., Calle de la Majada 3, 28760 Tres Cantos, Madrid, Spain; david.poudereux@altertechnology.com

*   Correspondence: manuel.c@upm.es or manuel.cano@inl.int; Tel.: +34910672661 or +351253140112 (ext. 2542)

Received: 27 March 2019; Accepted: 23 April 2019; Published: 26 April 2019

Abstract: In this work, the performance of a Mach–Zehnder interferometer manufactured in silicon nitride with a liquid crystal cladding is studied. The device consists of two multi-mode interference couplers linked by two rectangular waveguides, the cladding of one of which is a liquid crystal. The structure demonstrates the potential of using liquid crystals as tunable cladding material in simple waveguides as well as in more complex coupling or modulating structures. Liquid crystal cladding permits a local fine-tuning of the effective refractive index of the waveguide, avoiding coarse global temperature control. The study is realized in the visible region (632.8 nm), for its intrinsic interest region in (bio-)sensing or metrology.

Keywords: waveguides; polarization-selective devices; optical switching devices; optoelectronics

---

1. Introduction

Photonic integrated circuits, photonic chips or PICs, are increasingly being used in both optical communications and sensing. PICs consist of waveguides connecting different static or tunable elements such as filters and couplers, which convey several functionalities to the chip [1].

The manufacture of PIC waveguides conventionally relies on depositing and eliminating thin layers of material using photo-lithographic or electron beam processes. Often it is trivial to choose materials with the desired optical characteristics, e.g., silica and silicon nitride [2], which lead to cost-effective processes with good adherence between the different materials and acceptable yields. However, considering that electro-optically active elements mismatch the crystalline structure and affect all these features, compromises have to be made. Currently the most employed platforms are silicon-on-insulator (SOI) and indium phosphide (InP) [3]. InP allows the integration of passive and active (emitting/detecting) devices using a single platform technology [4], while SOI shares technology with complementary metal-oxide-semiconductors (CMOS) and has outstanding processing control, low-cost and high-volume processing.

Independently of which technology is employed, careful temperature control is often needed for the device to work as desired, since the thermal expansion of the chip and refractive index thermal dispersion affect the device behavior. In fact, local thermal control of elements distributed on a complex chip is a common [5] and relatively cost-effective way of tuning the elements, as recently illustrated in a reconfigurable photonic mesh [6]. However, thermal control has several drawbacks, its intrinsically difficult spatial isolation being the most important. Any heated element will over time transfer heat to neighboring elements; the closer the elements—i.e., the higher the degree of integration—the more severe the problem.

An alternative, similarly cost-efficient, tunable element could be a waveguide with liquid crystal (LC) cladding [7–10]. LCs are an intermediate state of matter between crystal solids and liquids. The simplest class of these materials, the thermotropic nematics, are typically viscous fluids of rod-shaped molecules [11]. These elongated molecules tend to arrange themselves in the same direction on a macroscopic scale, leading to fluids with both dielectric and optic uniaxial anisotropy. One may either consider the nematic LC as a fluid with orientational order, or as a crystal with no positional order.

Being fluid, the molecules tend to reorient with any applied field, even with relatively small applied voltages (a few volts per μm), and consequently so does the optical anisotropy (birefringence). Hence, by applying an electric field, one can reorient the LC optical axis (indicatrix) by modifying its optical properties. In most applications, one employs surface conditioning—an alignment layer—to induce a preferred initial—relaxed—orientation, and then applies an electrical field to reorient the liquid crystal. This easy procedure to modify the LC birefringence is the reason for LCs to be heavily applied to displays. In these devices, LCs are typically employed to modify the light state of polarization, allowing the light to be transmitted or not by placing the structure between crossed polarizers.

When substituting the conventional silica cladding of a waveguide with LC cladding, the device becomes electrooptically addressable. The effective refractive index of the cladding—as experienced by the light propagating in the core—becomes tunable between the LC ordinary and extraordinary indices (n_o and n_e) that affect the propagation of at least one of the polarization modes traveling in the waveguide (Figure 1).

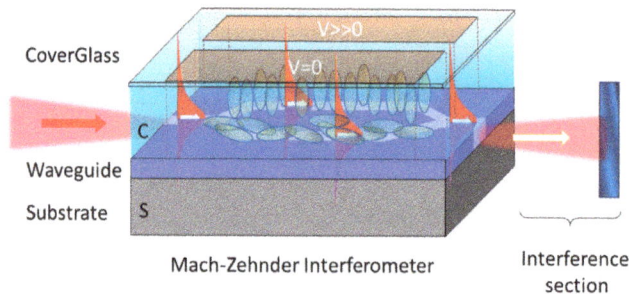

Figure 1. The principle of liquid crystal (LC) cladding tuning. An electric field is applied between the substrate and an external electrode. The effective refractive indices for the TE polarization in both arms of the Mach–Zehnder interferometer (MZI) are tunable. One arm is shown in the relaxed state (V = 0) and the other in a saturated state (V > V_{sat}). Any intermediate state would be possible.

A great advantage in using LC cladding tuning of the effective refractive index of waveguides is that by employing pixelated devices—like those of LC displays [12]—customized switching control is trivial. The typical pixel size in high-quality direct-view LC displays is about 100×100 μm^2, while in LC on silicon (LCoS) [13] displays for image projection, the pixel size is about 10×10 μm^2. In our facilities, we can adapt the pixel shape and size to any desired design. This allows us to address specific zones of the waveguide selectively. The electric field switching specific LC areas can be confined to μm-sized areas; the reduced thickness of the device avoids spreading of fringing. Hence, by employing pixelated LC devices, one can manufacture reconfigurable PICs with a density limited by the waveguide geometry rather than by the risk of thermal cross-talk between different components. A sandwich structure consisting of a custom designed waveguide structure, a thin LC layer and a generic or custom designed highly pixelated LC counter electrode would allow for the unprecedented complexity and density of the resulting PIC. Only one LC device could cover the entire PIC.

This work shows the first integrated Mach–Zehnder interferometer (MZI) driven by LC in visible (VIS) light. Integrated MZIs are the key tuning elements in the "optical field-programmable gate array (FPGA)" [6]. MZI LC structures have been presented before in fiber [14]; however, integrated devices are more stable than conventional or fiber MZIs to ambient thermal changes. Integrated MZIs are typically used as phase shifters, but they are becoming important modules (basic structures) for VIS light in (bio-)sensing or metrology. Moreover, considerable progress in nanophotonics has been focused on this spectral region [15,16].

2. Materials and Methods

The waveguide structure was designed and manufactured as a part of the Photonic Chip Design Training course (Valencia, Spain, 2016) [17]. It was fabricated by the laser direct write (LDW) technique. The platform is based on a 4" silicon wafer that is 500 µm thick, 2.5 µm thermal oxide bottom cladding, 300 nm silicon nitride core and 2 µm top cladding. A Heidelberg DWL 66FS was used for lithography writing. Deposition of the core was conducted by low pressure chemical vapor deposition (LPCVD). The top cladding was deposited by plasma-enhanced chemical vapor deposition (PECVD).

The integrated Mach–Zehnder interferometer (MZI) has a structure similar to any macroscopic MZI. The device consists of three sections; the first splits the input light power into two 50/50 channels. It is made of a 1 × 2 multimode interference (MMI) coupler with a length that couples exactly one half of the power into each output port (at 1550 nm). The second section consists of two parallel waveguides; differences in optical path length can be generated by changing the refractive index by heating [18] or electrically controlled tunable cladding. In the actual implementation, only one of the two interferometer arms is exposed to the tunable LC refractive index (Figure 2). The third section, where the interference arises, is a 2 × 2 MMI coupler that recombines light and sends it to one or the other output port depending on the phase delay difference between the arms. The two exiting waveguides are brought to the chip edge for monitoring. The waveguide dimensions were 0.3 µm × 0.6 µm, and the 1 × 2 MMI coupler and the 2 × 2 couplers were generated using the foundry recommendation for a wavelength of 1550 nm. The 50 µm curvature radii were used [19]. If SiO_2 cladding had been used, this design would ensure a single mode cut-off at 870 nm in the waveguides and dual mode behavior in the MMIs; however, when employing liquid crystal cladding, the cut-off depends on the switching state.

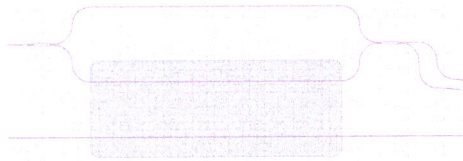

Figure 2. The tested MZI structure with multimode interference (MMI) in and out coupling. The LC is located in the dashed area. The straight line at the bottom is another waveguide that is not employed in this setup.

The 6 µm thick LC cell was made by sandwiching the LC material (Merck MDA-98-1602, n_o = 1.52, n_e = 1.78) [20] between the waveguide and an indium-tin-oxide (ITO) coated glass cover. The desired thickness was obtained using a 6 µm thick Mylar film between the chip and the glass cover.

In order to condition the alignment direction, the waveguide was gently rubbed with a velvet cloth, and the glass cover was spin-coated with polyimide [21] (PIA-2304, LixonAligner), which was likewise rubbed with the velvet cloth in the same direction.

Prior to assembly, the chip was mounted on a glass substrate for easy handling, the cover was precisely located and the cell was filled with the LC using capillarity. The sample was placed on a thermally stabilized XYZθ stage (Figure 3).

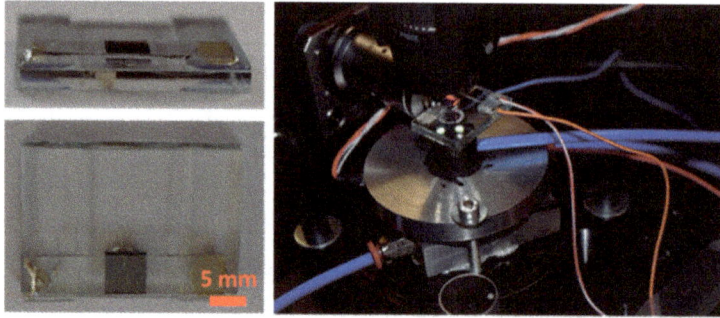

Figure 3. The mounted liquid crystal covered photonic chip. A HeNe laser was coupled to the MZI using a high-power microscope objective. The output was monitored using a Nikon D500 CMOS camera (Nikon Corporation, Shinagawa, Tokyo, Japan) with a macro lens.

3. Results and Discussion

The LC cell was excited using a 10 kHz square waveform with amplitude varying from 0 to 80 V. The LC alignment was set parallel to the waveguide. Figure 4 shows that the alignment was almost perfect.

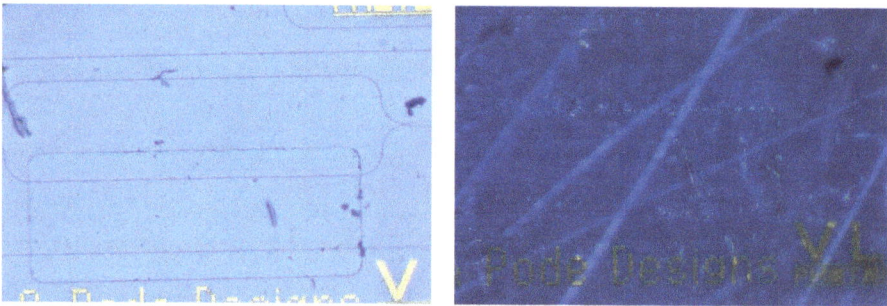

Figure 4. LC alignment as seen between crossed polarizers. The (**left**) image shows the clear state of the homogenous alignment with the sample aligned at 45° to the analyzer. In the (**right**) image, the sample is aligned with only a slight angle (3°) to the analyzer.

The variations of LC anisotropy upon reorientation are only experienced by the TM polarization perpendicular to the substrate. Therefore, only the TM mode is expected to vary its coupling ratio between outputs when the LC is switched. The transfer function for the output power of an MZI is $P_{out\ 1,2} = \frac{P_{in}}{2}(1 \pm cos\Delta\varphi)$, where the sub-indices 1, 2 are the different output ports and refer to the + and − signs in the equation; P_{in} is the input power.

Regarding the $\Delta\varphi$ as a function of voltage, it can be approximated as $\Delta\varphi = Ae^{B \cdot dc} + C$ [22], where dc is a duty cycle that is proportional to the voltage.

Figure 5 shows the light power at either output waveguide. The TE and TM modes were separated by placing a polarizer at the entrance of the MMI. In red colors (diamonds), one of the outputs (O1) of the MZI is represented. The blue colors (squares) represent the other output (O2). Due to significant noise levels, the data points were averaged over 10 repeated measurements. The errors are provided in vertical bars. The line is a simple spline fit to demonstrate the trend behavior.

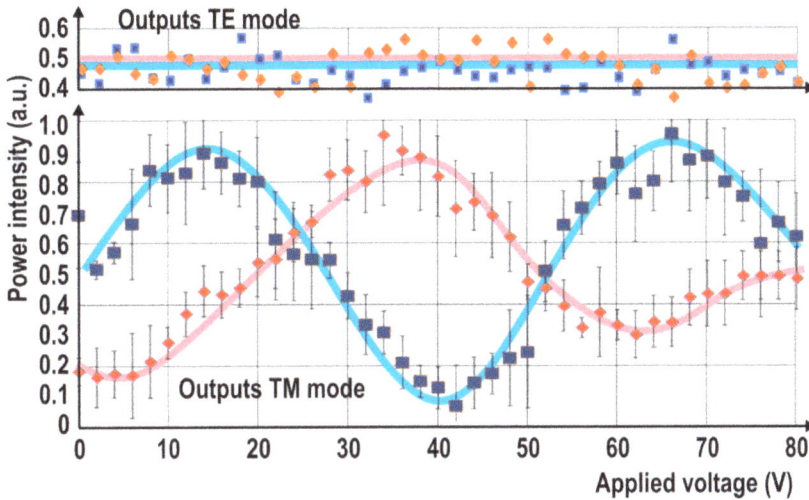

Figure 5. Transmission variation as a function of the applied field for TE (**top**) and TM (**bottom**) light modes. Only the TM mode is affected by the liquid crystal reorientation, since TE always experiences the LC ordinary index.

Like any MZI, the TM transmission shows a cosine-like variation $P_{out\ 1,2} = \frac{P_{in}}{2}(1 \pm cos\Delta\varphi)$. The induced phase variation depends, in a non-linear manner, on the effective refractive variation with the applied field, which can be described as $n_{LC} = Ae^{B \cdot V_{rms}} + C$ [22], and the change in degree of confinement of the traveling mode.

Obviously, the maxima and minima of the curves should match; a small offset between these is attributed to variation in the light coupling, or in scattering caused by the LC. TE transmission is substantially constant upon the whole driving voltage range, as shown by its nearly horizontal trend lines.

The MZI transfer function varies theoretically between zero and a maximum. In our case, residual light is obtained for every applied voltage. This can be clearly seen in Figure 6, where the maxima for either channel have been shown. In neither case is full extinction achieved.

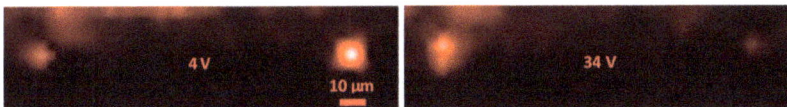

Figure 6. Typical output as captured by the camera. Clearly perfect extinction is not obtained when maximum intensity is transmitted, since the MMIs were designed for 1550 nm.

This is because the MMIs were designed for 1550 nm, as mentioned above, and not for the 632.8 nm working wavelength. Only by a perfect 50/50 power recombination in the output ports can full extinction be achieved. The designed MMIs led to a 63/37 power splitting for 632.8 nm light, leading to a theoretical extinction ratio of −5.6 dB. The discrepancy between the design and characterization wavelength is due to incompatibility between the available production development kits (PDK) necessary for the PIC production and the sample characterization tools available in our installations.

The relatively high voltages needed for full power switching were applied between the silicon substrate and the counter electrode. Hence, most of the voltage drop took place over the silicon-oxide and -nitride layers, rather than over the active LC layer. In future devices, the necessary voltages can be reduced by employing in-plane switching electrodes to a much more reasonable 3–5 V.

4. Conclusions

The results presented here in the VIS spectrum demonstrate that an LC-driven interferometer can be integrated in PICs. These results can be extrapolated to NIR. The LC birefringence generally reduces with increasing wavelength, but at the same time a longer wavelength will be less confined to the waveguide and therefore one should expect the NIR results to be similar, except for residual light. In comparison to MEMS, the pixelated LC offers the possibility of making complex phase devices at a very low price, albeit possibly at the cost of an increased insertion loss. As a standalone device, a switch based on LC cladding switching makes little sense.

The results show that the change in the LC switching state can be used for tuning the effective refractive index experienced by polarized light traveling in a waveguide. Hence, LC cladding may be employed as an alternative to thermal tuning, or in more involved applications as a substitute for conventional phase modulators such as LiNbO$_3$. Obviously, the use of LCs would be restricted to applications where response time is not an issue, as one would expect any nematic LC device to have response times in orders of tens of μs to tens or even hundreds of ms depending on the configurations.

Author Contributions: Design, M.C.-G. and D.P.; manufacturing, X.Q.; characterization, M.A.G., X.Q. and M.C.-G.; writing—review and editing, J.M.O., M.C.-G. and F.J.G.

Funding: This research was funded by the Spanish Ministerio de Economía y Competitividad (RETOS TEC2016-77242-C3-2-R, BES-2014-070964); Comunidad de Madrid and FEDER Program (S2018/NMT-4326).

Conflicts of Interest: The authors declare no conflict of interest.

References

1. Smit, M.K. Photonic Integrated Circuits. In Proceedings of the Integrated Photonics Research and Applications/Nanophotonics for Information Systems, San Diego, CA, USA, 10–13 April 2005.
2. Zaoui, W.S.; Kunze, A.; Vogel, W.; Berroth, M.; Butschke, J.; Letzkus, F.; Burghartz, J. Bridging the gap between optical fibers and silicon photonic integrated circuits. *Opt. Express OE* **2014**, *22*, 1277–1286. [CrossRef] [PubMed]
3. Liang, D.; Bowers, J.E. Photonic integration: Si or InP substrates? *Electro. Lett* **2009**, *45*, 578–581. [CrossRef]
4. Coldren, L.A.; Nicholes, S.C.; Johansson, L.; Ristic, S.; Guzzon, R.S.; Norberg, E.J.; Krishnamachari, U. High Performance InP-Based Photonic ICs–A Tutorial. *J. Light. Tech.* **2011**, *29*, 554–570. [CrossRef]
5. Orcutt, J.S.; Moss, B.; Sun, C.; Leu, J.; Georgas, M.; Shainline, J.; Zgraggen, E.; Li, H.; Sun, J.; Weaver, M.; et al. Open foundry platform for high-performance electronic-photonic integration. *Opt. Express OE* **2012**, *20*, 12222–12232. [CrossRef] [PubMed]
6. Pérez, D.; Gasulla, I.; Crudgington, L.; Thomson, D.J.; Khokhar, A.Z.; Li, K.; Cao, W.; Mashanovich, G.Z.; Capmany, J. Multipurpose silicon photonics signal processor core. *Nat. Commun.* **2017**, *8*, 636. [CrossRef] [PubMed]
7. Iv, J.B.; Hadeler, O.; Morris, S.M.; Wilkinson, T.D.; Penty, R.V.; White, I.H. Electro-Optic Integration of Liquid Crystal Cladding Switch with Multimode Passive Polymer Waveguides on PCB. In Proceedings of the Conference on Lasers and Electro-Optics/International Quantum Electronics Conference, Baltimore, MA, USA, 31 May–5 June 2009; p. CFV6.
8. Pfeifle, J.; Alloatti, L.; Freude, W.; Leuthold, J.; Koos, C. Silicon-organic hybrid phase shifter based on a slot waveguide with a liquid-crystal cladding. *Opt. Express OE* **2012**, *20*, 15359–15376. [CrossRef] [PubMed]
9. Xing, Y.; Ako, T.; George, J.P.; Korn, D.; Yu, H.; Verheyen, P.; Pantouvaki, M.; Lepage, G.; Absil, P.; Ruocco, A.; et al. Digitally Controlled Phase Shifter Using an SOI Slot Waveguide with Liquid Crystal Infiltration. *IEEE Photo. Tech. Lett.* **2015**, *27*, 1269–1272. [CrossRef]
10. Davis, S.R.; Farca, G.; Rommel, S.D.; Johnson, S.; Anderson, M.H. Liquid crystal waveguides: New devices enabled by >1000 waves of optical phase control. In Proceedings of the Emerging Liquid Crystal Technologies V, San Francisco, CA, USA, 25–27 January 2010; Volume 7618, p. 76180E.
11. Jeu, W.H.; de Jeu, W.H. *Physical Properties of Liquid Crystalline Materials*; CRC Press: Boca Raton, FL, USA, 1980; ISBN 978-0-677-04040-0.

12. Heilmeier, G.H.; Zanoni, L.A. Guest-host interactions in nematic liquid crystals. A new electro-optic effect. *Appl. Phys. Lett.* **1968**, *13*, 91–92. [CrossRef]

13. Zhang, Z.; You, Z.; Chu, D. Fundamentals of phase-only liquid crystal on silicon (LCOS) devices. *Light Sci. Appl.* **2014**, *3*, e213. [CrossRef]

14. Ho, B.-Y.; Su, H.-P.; Tseng, Y.-P.; Wu, S.-T.; Hwang, S.-J. Temperature effects of Mach-Zehnder interferometer using a liquid crystal-filled fiber. *Opt. Express OE* **2015**, *23*, 33588–33596. [CrossRef] [PubMed]

15. Maldonado, J.; González-Guerrero, A.B.; Domínguez, C.; Lechuga, L.M. Label-free bimodal waveguide immunosensor for rapid diagnosis of bacterial infections in cirrhotic patients. *Biosensors Bioelectr.* **2016**, *85*, 310–316. [CrossRef] [PubMed]

16. Muñoz, P.; Micó, G.; Bru, L.A.; Pastor, D.; Pérez, D.; Doménech, J.D.; Fernández, J.; Baños, R.; Gargallo, B.; Alemany, R.; et al. Silicon Nitride Photonic Integration Platforms for Visible, Near-Infrared and Mid-Infrared Applications. *Sensors* **2017**, *17*, 2088. [CrossRef] [PubMed]

17. Photonic Chip Design Training Course. Available online: http://www.vlcphotonics.com/ (accessed on 2 July 2018).

18. Li, J.; Gauza, S.; Wu, S.-T. Temperature effect on liquid crystal refractive indices. *J. Appl. Phys.* **2004**, *96*, 19–24. [CrossRef]

19. Muñoz, P.; Domenech, J.D.; Artundo, I.; den Bested, J.H.; Capmany, J. Evolution of fabless generic photonic integration. In Proceedings of the 2013 15th International Conference on Transparent Optical Networks (ICTON), Cartagena, Spain, 23–27 June 2013; pp. 1–3.

20. Caño-Garcia, M.; Elmogi, A.; Mattelin, M.-A.; Missinne, J.; Geday, M.A.; Oton, J.M.; Van Steenberge, G.; Quintana, X. All-organic switching polarizer based on polymer waveguides and liquid crystals. *Opt. Express* **2018**, *26*, 9584–9594. [CrossRef] [PubMed]

21. Kim, J.-H.; Yoneya, M.; Yamamoto, J.; Yokoyama, H. Nano-rubbing of a liquid crystal alignment layer by an atomic force microscope: A detailed characterization. *Nanotechnology* **2002**, *13*, 133. [CrossRef]

22. Caño-García, M.; Quintana, X.; Otón, J.M.; Geday, M.A. Dynamic multilevel spiral phase plate generator. *Sci. Rep.* **2018**, *8*, 15804. [CrossRef] [PubMed]

crystals

MDPI

Review

Liquid Crystal Beam Steering Devices: Principles, Recent Advances, and Future Developments

Ziqian He [1], Fangwang Gou [1], Ran Chen [1,2], Kun Yin [1], Tao Zhan [1] and Shin-Tson Wu [1,*]

[1] College of Optics and Photonics, University of Central Florida, Orlando, Florida 32816, USA;
zhe@knights.ucf.edu (Z.H.); fangwang.gou@knights.ucf.edu (F.G.); tradchenr@knights.ucf.edu (R.C.);
kunyin@knights.ucf.edu (K.Y.); tao.zhan@knights.ucf.edu (T.Z.)

[2] Key Laboratory of Applied Surface and Colloid Chemistry, School of Materials Science and Engineering,
Shaanxi Normal University, Xi'an 710119, China

* Correspondence: swu@creol.ucf.edu; Tel.: +1-407-823-4763

Received: 9 May 2019; Accepted: 3 June 2019; Published: 5 June 2019

Abstract: Continuous, wide field-of-view, high-efficiency, and fast-response beam steering devices are desirable in a plethora of applications. Liquid crystals (LCs)—soft, bi-refringent, and self-assembled materials which respond to various external stimuli—are especially promising for fulfilling these demands. In this paper, we review recent advances in LC beam steering devices. We first describe the general operation principles of LC beam steering techniques. Next, we delve into different kinds of beam steering devices, compare their pros and cons, and propose a new LC-cladding waveguide beam steerer using resistive electrodes and present our simulation results. Finally, two future development challenges are addressed: Fast response time for mid-wave infrared (MWIR) beam steering, and device hybridization for large-angle, high-efficiency, and continuous beam steering. To achieve fast response times for MWIR beam steering using a transmission-type optical phased array, we develop a low-loss polymer-network liquid crystal and characterize its electro-optical properties.

Keywords: liquid crystals; beam steering; optical phased arrays; liquid-crystal waveguides; Pancharatnam-Berry phase; volume gratings; fast response time

---

1. Introduction

Precisely positioning a laser beam or light ray is crucial for practical applications, such as light detection and ranging (LiDAR) [1–3], displays [4–6], microscopy [7], optical tweezers [8], and laser micro-machining [9]. For example, LiDAR, as the most pivotal application promoting beam steering techniques, can map landscapes in a three-dimensional (3D) space and serves as an enabling technology for space station navigation, telescope docking, and autonomous cars, drones, and underwater vehicles [10–12]. To date, a plethora of beam steering methods have been demonstrated. Generally, they can be divided into two groups: Mechanical and non-mechanical beam controls. Mechanical approaches include scanning/rotating mirrors [13], rotating prisms [14], piezo actuators [15], and micro-electromechanical system (MEMS) mirrors [16,17]. On the other hand, non-mechanical options include acousto-optic and electro-optic deflectors [18–21], electro-wetting [22–25], and liquid crystal (LC) technologies [26,27], to name a few. Although traditional mechanical beam steering devices are reasonably robust, some technical issues remain to be overcome, such as relatively short lifetimes, heavy weight, large power consumption, and high cost. In contrast, recently-developed mechanical and non-mechanical beam steerers show promise for resolving these shortcomings [28]. As a strong candidate, LC-based beam steerers can be lightweight, compact, consume low amounts of power, and inexpensive.

LCs are self-assembled soft materials, consisting of certain anisotropic molecules with orientational orders. They can respond to various external stimuli, including heat, electric and magnetic fields, and light [29–32]. For instance, in the presence of an electric field, LC directors can be re-oriented, due to both the optical and dielectric anisotropies of the LC molecules, resulting in refractive index modulation (bi-refringence). Using this simple principle, LC spatial light modulators (SLMs)—also called LC optical phased arrays (OPAs)—can be established by pixelating such refractive index modulators in a two-dimensional (2D) array [33]. While LC-based OPAs were developed more than three decades ago, they continue to advance. Meanwhile, other LC-based beam steerers, such as compound prisms, resistive electrodes, LC-cladding waveguides, Pancharatnam-Berry phase deflectors, and LC volume gratings, have also emerged, exhibiting great potential for new applications.

In this paper, we review recent advances in LC-based beam steering device technology. In Section 2, we describe the physical principles of LC-based beam steering. In Section 3, we briefly introduce different types of LC-based beam steerers and summarize their pros and cons, and then focus on more recently-developed devices and technologies. In addition, we propose an improved LC-cladding waveguide beam steerer. In Section 4, we emphasize two future technical challenges: Fast response times, especially for long-wavelength beam steering, and device hybridization for large-angle, high-efficiency continuous beam steering. With our low-loss polymer-network liquid crystal, a transmission-type LC phase modulator with a relatively fast response time and workable operation voltage can be achieved for mid-wave infrared (MWIR) applications.

2. Operation Principles of LC Beam Steering

Several LC-based beam steering devices have been developed, and their basic operation principles can be classified into three major categories: Blazed gratings (Raman-Nath diffraction), Bragg gratings (Bragg reflection), and prisms (refraction). Of course, other mechanisms exist which can also lead to high performance beam steering. One such example is cascaded lenses. Previous works have shown that, by stacking liquid lenses in a de-centered manner, large steering angles can be achieved [34]. This concept may apply to LC lenses, as well. However, LC lenses usually have a limited tuning range for optical power and are polarization-dependent. Another example is cascaded micro-lens arrays [28]. Traditionally, beam steering can be fulfilled by moving the micro-lens arrays away from the center (optical axis). By tuning the curvature of the LC micro-lens, it is possible to have a beam steering system without mechanical movement. Although refraction-type LC micro-lens technology has advanced greatly, the major concerns are the relatively slow response time (approximately 100 ms) and limited optical dynamic range; while switchable diffractive lenses do not present much advantages over the traditional ones, and may still need mechanical parts [35,36].

Here, we make clear that the research into LC-based beam steering is still ongoing. Apart from the above-mentioned options, other scientifically intriguing approaches have also emerged [37,38]. For instance, by combining micro-mirrors with LC elastomer fibers, it is possible to create a beam steering device which responds to external stimuli, such as light [37]. These methods are still in under development and are not ready for practical applications, but may find interesting applications in the future.

2.1. Blazed Gratings

Figure 1 depicts the schematic of a one-dimensional (1D) grating. Generally, for a thin grating (Raman-Nath regime) [39], an incident beam can result in multiple diffracted beams, in both backward (reflection) and forward (transmission) directions. The diffraction angles are determined by the following grating equation [40]:

$$n_{(1,2)} \sin \theta_{m,(1,2)} = n_1 \sin \theta_{in} - m\frac{\lambda_0}{\Lambda} \qquad (1)$$

where $n_{(1,2)}$ is the refractive index in medium 1 or 2, $\theta_{m,(1,2)}$ is the angle of *m*-th diffraction order in medium 1 or 2, θ_{in} is the incident angle, Λ is the grating period, and λ_0 is the wavelength in vacuum.

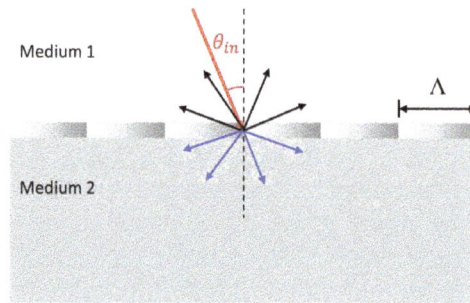

Figure 1. Schematic drawing of a grating in the Raman-Nath regime. Multiple diffraction orders in both the forward (blue arrows) and backward (black arrows) directions coexist.

A desirable feature of the blazed grating is that a designed diffraction order can achieve maximum efficiency. In particular, the phase profile of a plane-wave passing through a blazed grating will become sawtooth-like. As depicted in Figure 2, there is a phase reset after each grating period. To reach maximum efficiency, for a given order m, the maximum phase difference within one period (or phase reset value) needs to be $2m\pi$. Based on the highly selective diffraction of blazed gratings, two beam steering approaches can be realized. The first is to vary the grating period: As shown in Figure 2a, as long as it stays at the blaze condition, maximum diffraction efficiency can be maintained at the m-th order (normally, 1st order) while θ_m changes according to the grating period. In this approach, the steering angle can be tuned continuously, in principle. The second approach is through varying the phase reset value—for example, from p-th order to q-th order (or other orders)—as Figure 2b shows. As the grating period is fixed, the steering angle can only achieve discrete values, as described by Equation (1).

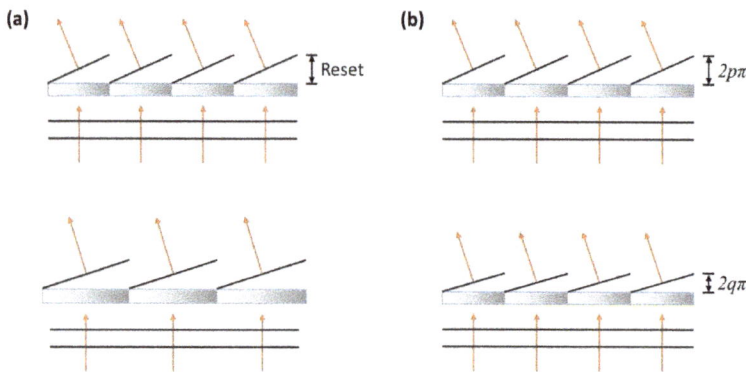

Figure 2. Illustration of beam steering based on (**a**) variable-period and (**b**) variable diffraction order blazed gratings.

2.2. Bragg Gratings

The second category is volume grating. Unlike the above-mentioned blazed grating, volume gratings are based on thick gratings (Bragg regime). Figure 3a depicts a 1D Bragg grating, where the blue lines are the periodic planes. In such a thick grating, the establishment of high-efficiency diffraction needs several periods, and the reflections from each plane add up coherently. A distinct feature of these gratings is a high sensitivity to incident wavelengths and angles. The Bragg condition can be described as:

$$2d \sin \theta = m\lambda \tag{2}$$

where d is the period, θ is the angle between planes and incident beams, and λ is the wavelength in the medium. A well-known example of LC Bragg gratings is cholesteric liquid crystal (CLC), as schematically illustrated in Figure 3b. A CLC structure can be formed by doping some chiral compounds into a rod-like nematic host. In a CLC phase, the LC directors rotate continuously along the stacking direction, and a helical structure perpendicular to the layer planes (orange dashed lines) exists. To establish Bragg reflection, usually about 10 pitches are needed [29]. The reflection is also strongly dependent on the incident angle and wavelength. By dynamically tuning the pitch length, the angle, as well as wavelength dependency, changes accordingly [41]. This is the principle employed for volume grating beam steering: By changing the period, within the bandwidth, the diffraction angle for a given wavelength varies. Another distinct feature is that a thick grating can steer incident light at a large angle with high efficiency, while a thin grating can hardly achieve that [42].

(a)　　　　　　　　　　　　　**(b)**

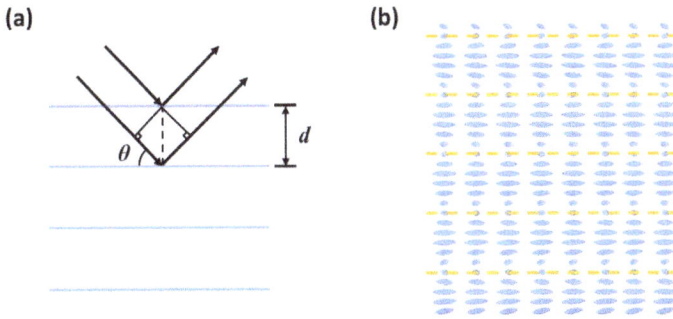

Figure 3. (a) Schematic sketch of Bragg reflection in a periodic structure, and (b) a liquid crystal (LC) Bragg grating based on a helical cholesteric structure.

2.3. Prism-Type Beam Steering

The third mechanism is prismatic beam steering. Differing from the above-mentioned blazed and volume gratings, prismatic beam steering is based on refraction rather than diffraction. As shown in Figure 4, when a light beam passes through a prism, the refraction at both interfaces leads to a deviated outgoing beam. Through a geometric optics calculation, the deviation angle δ can be expressed as:

$$\delta = \theta_{in} + \sin^{-1}\left(\sin\phi\sqrt{\frac{n_2^2}{n_1^2} - \sin^2\theta_{in}} - \cos\phi\sin\theta_{in}\right) - \phi \tag{3}$$

where ϕ denotes the prism angle, n_1 is the refractive index of background medium, and n_2 is the refractive index of the prism. When the prism angle and the incident angle are small, Equation 3 can be simplified to:

$$\delta \approx \left(\frac{n_2}{n_1} - 1\right)\phi \tag{4}$$

For such a thin prism, the angular deviation is insensitive to the angle of incidence (θ_{in}) over a decent range, but it depends on the refractive indices of both media (n_1 and n_2) and the prism angle (φ). By this principle, we can tune the deviation angle by changing one of the refractive indices. However, traditional LC prisms can only vary in deviation angle over a small range, due to limited interaction lengths [43]. As will be discussed in Section 3.4, by increasing the interaction length, it is possible to obtain a large deviation angle (even up to 270°) through this mechanism.

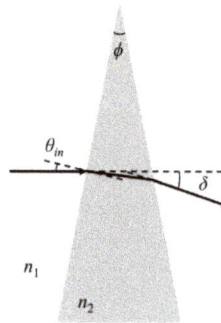

Figure 4. Schematic diagram of a prismatic beam deflector; δ is the deviation angle.

3. Liquid Crystal Beam Steering Devices

In Section 2, we discussed the basic operation mechanisms of LC beam steering devices. In this section, we focus on six device configurations, including optical phased arrays, compound blazed gratings, resistive electrodes, LC-cladding waveguides, Pancharatnam–Berry phase gratings, and volume gratings. For each device, we will discuss its merits and demerits, and describe to which mechanism it belongs.

3.1. Optical Phased Arrays

The LC-based OPA is a relatively mature beam steering technology, where homogeneous alignment with positive dielectric anisotropy (Δε) LC materials or vertical alignment with negative Δε LC materials can be considered. However, homogeneous alignment with a positive Δε LC is preferred, as negative Δε LC materials usually show a higher viscosity and lower |Δε|, leading to a slower response time and higher operation voltage [44]. These conventional OPAs intrinsically work only for one linear polarization and some design considerations have already been thoroughly discussed in several papers [28,33,45]. The mechanism of this type is a variable-period blazed grating (Figure 2a) with a 2π phase reset (only the 1st order is applied). Currently, the main issue for OPAs (of both homogeneous and vertical alignments) is the fringing field effect, which occurs when different voltages are applied on adjacent pixels, resulting in an unwanted phase distortion to the LC directors. This distortion becomes more noticeable as the pixel pitch to cell gap ratio gets smaller. For instance, for a reflection-type OPA—such as LC-on-Silicon (LCoS)—the cell gap can be reduced to one-half of the transmission type while keeping the same phase change. Under such conditions, the fringing field effect will be smaller. The most severe effect appears at the phase reset region, which is often referred to as the flyback region [46], as depicted in Figure 5. The efficiency drop due to the flyback region can be estimated by [33]:

$$\eta = \left(1 - \frac{w}{\Lambda}\right)^2 \tag{5}$$

where η is the efficiency, w is the width of the flyback region, and Λ is the width between phase resets. From Equation 5, for a fixed w, efficiency drops drastically with decreasing period; that is, the OPA can only maintain a relatively high efficiency within a small steering angle range (e.g., at ±5°, the optical efficiency has already dropped to approximately 60%). The beneficial part is that, within this small tuning range, the steering is quasi-continuous. OPA works well in the visible and near-infrared spectral regions. However, to extend its usable range to mid-wave infrared (MWIR), a thicker LC layer is needed in order to accumulate a 2π phase modulo, which will lead to a slow response time. To address this issue, we propose an improved LC mixture; detailed results will be discussed in Section 4.1.

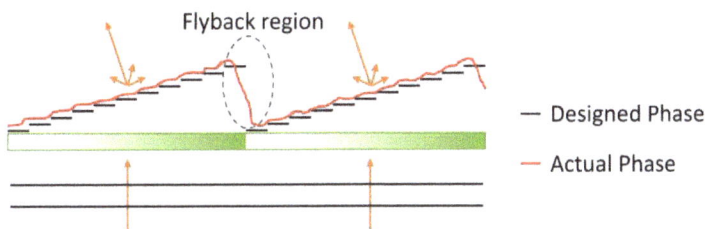

Figure 5. After passing through the optical phase array (OPA), the designed step-wise phase profile is distorted, especially in the flyback regions.

3.2. Compound Blazed Gratings

The compound blazed grating was first demonstrated almost two decades ago [47]. Utilizing electron beam lithography, a PMMA (Poly(methyl methacrylate)) blazed grating can be fabricated on top of an indium tin oxide (ITO)-glass substrate. The final device is accomplished through the assembly of a homogeneously-aligned LC cell and the diffraction appears only for one linear polarization parallel to the LC alignment direction. The working mechanism is the variable diffraction order blazed grating (Figure 2b), where the devices are switchable between the 1st (voltage-off) and 0th (voltage-on) orders. By matching the refractive index of the PMMA with the ordinary refractive index of the employed LC, unwanted diffractions in the voltage-on state can be eliminated. In [47], four devices with different grating periods were fabricated and stacked together to realize multi-angle beam steering. The efficiencies of steering a beam to 13.5° (largest angle) and 0° were approximately 30% and 58%, respectively. Beside the Fresnel loss by stacking multiple devices, the device efficiency mainly suffered from the flyback effect, caused by the imperfect PMMA blazed grating fabrication and distorted LC directors. Although current e-beam lithography systems can achieve better control, the fabrication method itself is not a low-cost choice.

To lower the cost of the sophisticated fabrication process, we could consider the imprinting method, which has exhibited good reliability. The fabricated PMMA blazed gratings are much larger, such that applying the same mechanism (variable diffraction order blazed grating), the imprinted blazed gratings can switch among higher orders. Two configurations have been proposed; namely, the transmissive [48] and reflective [49] modes, as shown in Figure 6a,b, respectively. In transmissive mode, the voltage-off state provides the highest steering angle. As the voltage increases, the steering angle starts to decrease (for example, from 4th order to 3rd order), approaching even negative orders (−1st), and jumps back to the 0th order. The reason it reaches negative orders is that the dielectric constants of the LC materials (E7, in this case) and PMMA are different. The voltage shielding effect in the thin PMMA region is smaller than that in the thick region. As a result, at a given voltage, the LC directors in the thin PMMA regions are already tilted up, but they are not fully reoriented in the thick PMMA regions. This is an important aspect to take into consideration when designing the compound structures. On the other hand, a reflective mode can be generated by simply depositing a thin reflective film (Ag, Al, or so on) on top of the PMMA. It is worth mentioning that, in the reflective mode, even in the state where the LCs are all tilted up, the light beam will be directed to a large diffraction order, possibly achieving large angles (e.g., 30°), rather than the 0th order, due to the optical path difference introduced by reflection. For both configurations, the steering angle is discrete and the steering range is limited to several degrees. The steering efficiency can approach 100%, in theory, if an ideal blazed phase profile is satisfied. However, experimentally, the efficiency is only 30–50%; this is mainly due to the flyback effect, but other factors, such as Fresnel reflections and disordered LC directors, also contribute to the efficiency loss.

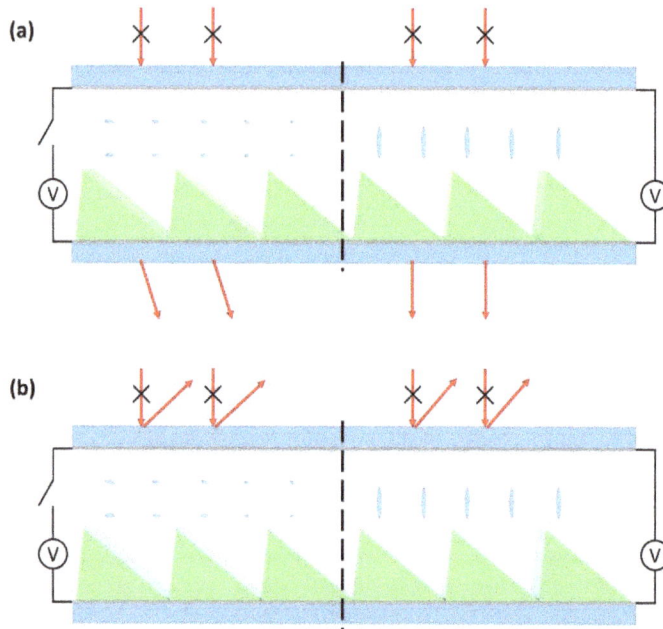

Figure 6. Two device configurations of compound blazed gratings: (**a**) Transmissive mode and (**b**) reflective mode. At a large applied voltage, the transmissive mode will direct light to small diffraction orders, while the reflective one will guide light to large diffraction orders. The incident linear polarization is parallel to the LC alignment direction.

3.3. Resistive Electrodes

Beam steering using resistive electrodes can be realized by creating a gradient voltage drop in a homogeneous or vertical alignment LC cell. Figure 7 depicts the device configuration of such a LC cell, with a resistive electrode on the top substrate. Overall, there are three types of electrode with different functions. The planar common electrodes at the bottom substrate are grounded, while the conductive electrodes create a voltage difference across the resistive electrodes (usually a resistive film). The working mechanism is the variable diffraction order blazed grating and the maximum diffraction angle depends on the electrode dimensions and LC layer thickness. A beam steering device using metal for conductive electrodes and a metal-oxide for resistive electrodes was proposed and fabricated about two decades ago [50]. A 76% diffraction efficiency was obtained within the maximum deflection range of ±0.34°. More recently, employing patterned ITO for the conductive electrodes and poly(3, 4-ethylenedioxythiophene):polystyrene sulfonate (PEDOT:PSS) for the resistive electrodes, a maximum steering angle of 4.8° was achieved, where the steering angle could be varied among many diffraction orders [51]. However, due to the thick cell gap (about 30 μm), the response time was slow (several seconds); further, the efficiency was not very high, as the steered beam spots were gloomy (the exact efficiency was not reported). This approach mainly suffered from the flyback region effect and a non-ideal phase gradient, along with other concerns with regards to increased power consumption resulting from the resistive electrodes. Nevertheless, to improve the phase gradient, a device optimization method has been developed [52].

In fact, to achieve a phase gradient without compound structures, the resistive electrode approach is not the only option. For instance, by step-by-step polymerization of reactive mesogens in a homogeneous alignment cell, phase-gradient gratings can be generated where the phase profile is controlled by applying different voltages [53,54]. Nonetheless, the fabrication process is complicated, especially for small steps, and defects could form during polymerization, which will ultimately limit the performance of the device.

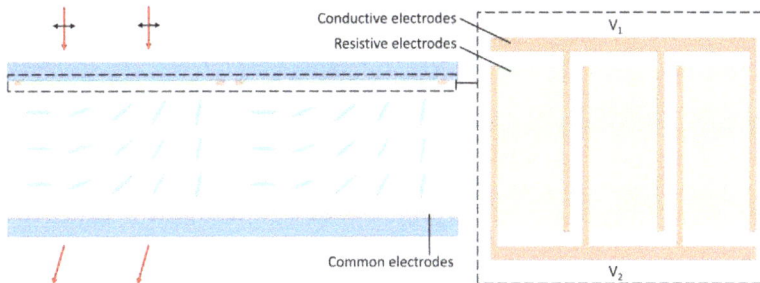

Figure 7. Layout of the resistive electrode-based beam steerer. The dashed rectangle depicts the top view of the top electrodes.

3.4. LC-Cladding Waveguides

As mentioned in Section 2, LC prisms can only achieve relatively small steering angles, due to the limitations in LC refractive index change and prism geometry [43]. However, the interaction length can be dramatically increased by guiding light using a waveguide, as Vescent Photonics demonstrated about a decade ago [55]. The main feature of in-plane steering is plotted in Figure 8. This special waveguide consists of LC cladding, a Si_3N_4 core, and glass cladding. Beam steering is achieved by manipulating the evanescent wave in the LC cladding region. The homogeneous LC alignment is along the light propagation direction, and the top zigzag-shaped electrodes are divided into two parts. If a voltage is applied on the steer-up electrode while the other electrodes are grounded, the transverse-magnetic (TM) mode will experience multiple prisms, where the region under the steer-up electrode has a higher refractive index. According to Equations 3 and 4, the light will be steered up. By controlling the applied voltage, the refractive index contrast of the prisms can be managed. Therefore, this configuration has a very high angular resolution. As the LC prisms only interact with evanescent waves, only the LC directors near the interface are involved. Thus, the response time is in sub-millisecond range. By special design of the electrodes, a very large steering angle (up to 270°) has been demonstrated. This device can also steer light in the vertical dimension with a smaller range (for example, from 0° to 15°), achieved by adding another set of electrodes and using the tunneling mode. The major losses are from the Fresnel loss upon refractions and some in-coupling and out-coupling losses. However, as demonstrated, a waveguide amplifier can compensate for the losses [56]. The aforementioned systems work for near-infrared and short-wave infrared. Some follow-up works have extended the beam steering device into the MWIR range [57] with optimized materials that minimize absorption losses.

The LC-cladding waveguide is a general approach to extend the interaction length between LCs and light beams. As mentioned in Section 3.3, the resistive electrode approach has the same concern as the LC prisms in that, to increase the steering angle, a thick LC layer is needed. Here, we propose a beam steering concept that combines the LC-cladding waveguide with resistive electrodes. The device configuration is shown in Figure 9. In fact, the structure is similar to the previous waveguide beam steerer, except that the top electrodes are replaced by resistive electrodes. By creating a refractive index gradient and interacting with the evanescent waves, the LC thickness can be thin, and the accumulated phase gradient can be large, meaning that large-angle beam steering with fast response time is achievable. Ideally, a single resistive electrode (structure I in Figure 9) can provide continuous beam steering, as it is refractive. However, the width of the resistive electrode is closely related to the refractive index gradient and, thus, to the maximum steering angle (if the total length of the LC-cladding waveguide is fixed). For a compact design, a limited width of the resistive electrode is preferred, but the trade-off is that it can only work for small-sized beams. To enlarge the aperture size while maintaining the same maximum steering angle, a wider resistive electrode with a longer LC-cladding waveguide or an array of resistive electrodes can be applied (structure II in Figure 9). When designing the width of a single resistive electrode, we also need to consider the sheet resistance. Here, we perform finite-difference time-domain (FDTD) simulations to prove this concept. Due to the dimension limit in the simulations, we only test the performance of structure I. The parameters employed in our simulation are as listed below, unless otherwise stated; 1) The glass cladding material is SiO_2; 2) the core material is Si_3N_4, with thickness of 500 nm and refractive index 2.0; 3) the LC alignment layer on the core is 20 nm thick, with a refractive index of 1.5; 4) the LC material has an ordinary refractive index of 1.5 and an extraordinary refractive index of 1.75; 5) the homogeneous alignment direction is parallel to the waveguide direction; 6) the width of the resistive electrodes is 40 μm for case 1 and 60 μm for case 2; 7) the waveguide length is 150 μm; and 8) the incident wavelength is $\lambda = 1.55$ μm and the incident beam is of fundamental TM mode.

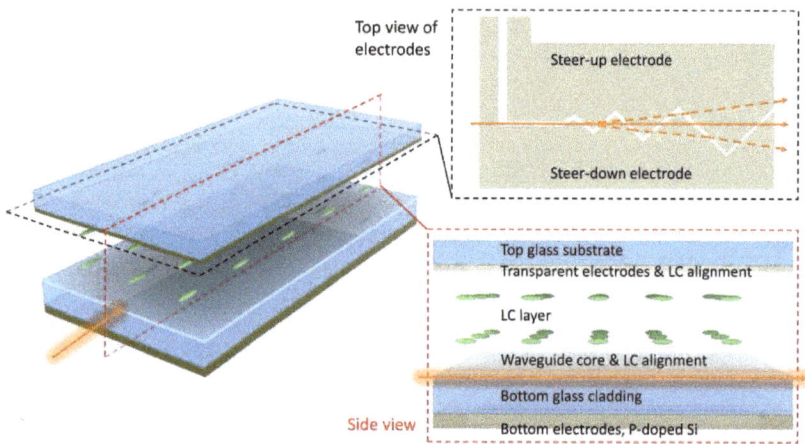

Figure 8. Schematic plot of the LC-cladding waveguide beam steerer, proposed by Vescent Photonics.

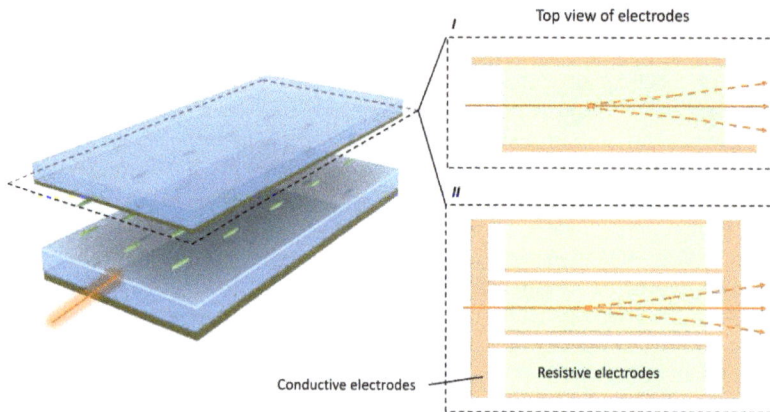

Figure 9. Schematic drawing of the LC-cladding waveguide beam steerer based on resistive electrodes. The major difference from the Vescent Photonics device is the structure of the top electrodes, where structure I uses a single resistive electrode and structure II uses an array of resistive electrodes.

During the simulations, we analyzed two cases (1 and 2), with different refractive index gradients, by defining various refractive index differences (Δn_w) between the two conductive electrodes and assuming that the refractive index changes linearly across the resistive electrode. In our case, the maximum bi-refringence was assumed to be Δn = 0.25. Although previous reports have demonstrated even larger Δn at 1.55 µm, the parasitic large visco-elastic constant will strongly increase the device response time and operating voltage [58,59]. Figure 10 depicts the simulated results (steering angle versus refractive index difference). The steering angle monotonically increases with the refractive index difference and the slope of increase is higher for a more compact dimension (case 1). For case 1 (40 µm resistive electrode width), a maximum steering angle of approximately 4.9° can be obtained within only 150 µm of propagation length. Compared to the prism-type waveguide, the resistive electrode-type waveguide could achieve the same steering angle at a shorter waveguiding distance, as the interaction between LC and light is continuous, while the interaction for the prism-type waveguide is discrete (only at the prism boundaries). However, the tradeoff is that, if structure II is utilized for a larger-size beam, the flyback region effect will again limit the diffraction efficiency. It should also be pointed out that the simulations are based on rather ideal situations, without an electrical response simulation of the LC directors. In the actual case, the refractive index change is not ideally linear and the LCs near the alignment layer should be less tilted due to the strong anchoring effect, which results in somewhat lower steering angles and efficiency. However, the refractive index change across the resistive electrode can be further optimized and the surface anchoring of LC directors can be effectively simulated by increasing the alignment layer thickness.

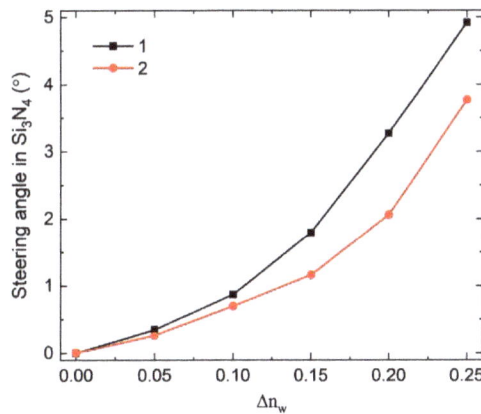

Figure 10. Simulation results of steering angle versus refractive index differences for both case 1 (black) and case 2 (red).

3.5. Pancharatnam-Berry Phase Deflecctors

With intrinsic polarization sensitivity and high diffraction efficiency, Pancharatnam-Berry deflectors (PBDs) [60,61], made of liquid crystals or liquid crystal polymers, are another promising candidate for non-mechanical beam steering. Unlike the traditional dynamic phase produced through optical path difference, the Pancharatnam-Berry (PB) phase corresponds to the phase shift introduced by other light-wave parameters, such as polarization transformation [62,63]. As a special case of PB optical elements [64–66], PBDs can be established by patterning half-wave plates with an in-plane linearly rotating optical axis, as depicted in Figure 11a. Their working mechanism can be simply explained by the Jones matrix:

$$R(-\varphi)W(\pi)R(\varphi)J_\pm = -je^{\pm 2j\varphi}J_\mp \tag{6}$$

where R is the rotation matrix, φ is the optical axis orientation angle, $W(\pi)$ is the half-wave retardation matrix, and $J_\pm$ are the Jones vectors of right-handed circularly polarized (RCP) or left-handed circularly polarized input light (LCP). With a linearly-rotating optical axis $\varphi = \pi x/p$, where p is the grating period, the PBD can deflect circularly polarized normal incident light into the first order with near 100% diffraction efficiency, in theory (within paraxial approximation), and flip the handedness of the light. Nonetheless, there are some limitations. The diffraction efficiency can reach 100% only within a relatively small angle, depending on the bi-refringence of the employed LC material (e.g., approximately 15^0 with Δn about 0.2) [64]. Another study has demonstrated that the LCs cannot follow periodic alignment patterns well when the cell gap is less than a single period, where the LC directors rotate 180^0 in the plane [67]. To fabricate PBDs, polarization holography [68,69] and photo-alignment methods [70] are usually applied to transfer the linearly rotating polarization pattern to the LC director orientation. In terms of the driving method, PBDs can be categorized into two types: Passive and active. Passive PBDs are generally prepared by coating a reactive mesogen layer onto an aligned surface (Figure 11a), while active PBDs are fabricated by filling LC into a cell with transparent electrodes and a patterned photo-alignment layer (Figure 11b). If the applied voltage is high enough (usually <15 V), the LC directors are re-oriented along the electric field directions, so that the deflection effect of the PBD vanishes.

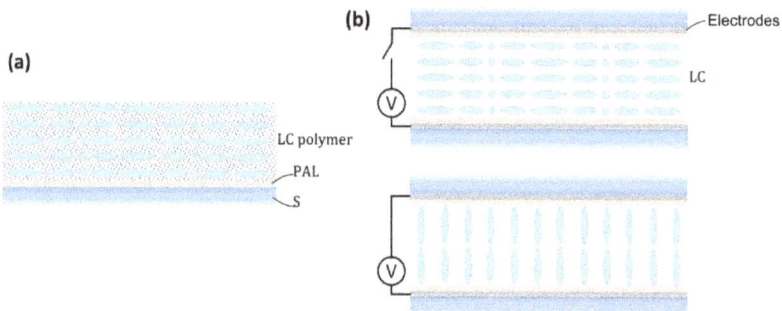

Figure 11. Schematic illustration of the structure of (**a**) a passive Pancharatnam-Berry deflector (PBD) made of a liquid crystal polymer, and (**b**) an active PBD made of liquid crystal at voltage-off and voltage-on states. PAL, photo-alignment layer; S, substrate.

In principle, two beam steering schemes can be realized through PBDs. Firstly, because the diffraction is highly polarization-selective, the deflection angle of the PBD can be switched by controlling the input polarization state, such that the output beam can switch between +1 and −1 diffraction orders, as depicted in Figure 12a. In this scheme, a polarization rotator (e.g., a 90° twisted-nematic cell) and a quarter-wave plate are placed in front of the PBD to dynamically select the input polarization. Secondly, for active PBDs made of liquid crystals, the input beam can be switched between the zeroth and first diffraction orders, simply by applying a voltage to the PBD, as illustrated in Figure 12b. The working principles are that of the variable diffraction order blazed grating and the most intriguing property is that these devices are free from the flyback issue. To obtain multiple steering angles, the simplest way is to stack multiple PBDs together [71]. With N-cascaded PBDs working as illustrated in Figure 12a,b, beam steering with 2^N discrete angles can be realized. Apart from the binary switch through the aforementioned two approaches, a ternary switch is also possible if we cascade N PBDs that are both switchable and with polarization rotators and quarter-wave plates (combining both approaches) [72]; then, 3^N discrete angles can be generated. Furthermore, thanks to the diversity of LC and LC polymer materials, PBD-based beam steering devices have been demonstrated in the visible [73], near infrared [72], and MWIR [74] regions. More recently, multi-twist structures [75] and twisted-nematic diffractive waveplates [76,77] have been proposed and were demonstrated to show a broader bandwidth (from visible to infrared), in comparison with traditional PBDs. Furthermore, 2D PBDs [78–81], which can diffract beams into multiple directions, have recently emerged. In addition to varying the diffraction orders, there have been other proposals to change the period by using a sophisticated design of the alignment layer and electrodes [82]. However, the feasibility of this approach remains to be proven.

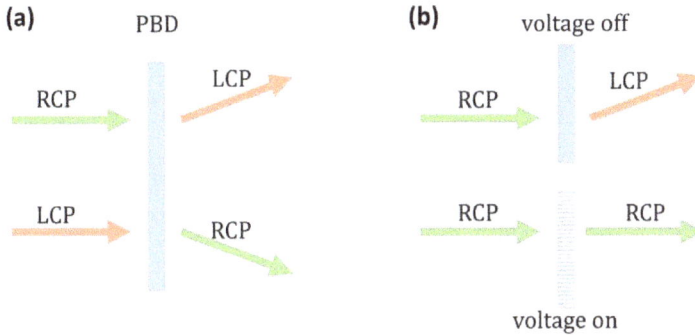

Figure 12. Schematic illustration of the two beam steering approaches using PBD: (**a**) Input polarization switch (i.e., a passive switch), and (**b**) PBD structure switching between on and off states (i.e., an active switch).

3.6. LC Volume Gratings

Volume gratings (Bragg regime) have the ability to split and diffract beams into different directions. With meticulous design, volume gratings exhibit several distinctive features: Strong polarization selectivity, high diffraction efficiency, and large deflection angles. As a typical volume grating, the early-developed holographic polymer-dispersed LC (HPDLC) is a candidate for beam steering applications with switching ability and polarization dependency [83,84]. By stacking multiple holographic volume gratings together, discrete steering angles can be realized. However, fabrication challenges exist for high quality HPDLC-based holographic gratings, due to the induced inhomogeneity during the diffusion process, resulting in optical imperfections in the structure [84].

Besides HPDLC, the recently-developed LC polarization volume gratings (PVGs) are more attractive, due to their near-100% diffraction efficiency [42,85–90]. Figure 13a illustrates the LC director distribution of a reflective PVG [85]. The Bragg period and periods along the x and y directions are Λ_B, Λ_x, and Λ_y, respectively. This asymmetrically self-organized structure enables unique polarization selectivity, where only light with the same chirality as the helical twist will be diffracted. Figure 13b depicts the polarization states of the diffracted and transmitted beams for a typical reflective PVG. Here, we assume that the handedness of the PVG is right-handed along the incident direction. Therefore, such a PVG can diffract the RCP light to the first order and keep its polarization state. Meanwhile, the LCP light will pass through the PVG without changing its polarization state. The same working principles can be applied to transmissive PVGs, except that the diffracted light will flip its handedness [42]. For instance, if an RCP light is incident on a transmissive right-handed PVG, the outgoing light will be diffracted into the first order with polarization changed to LCP.

Compared to other LC-based devices (such as PBDs), PVGs possess a larger diffraction angle with high efficiency. With proper organization, the diffraction angle can reach 70° or even larger, while keeping diffraction efficiency over 90% for a single PVG. These unique advantages make PVG a promising candidate for beam steering applications. While PVGs can be applied as a passive beam steerer, with the help of the fast-switching polarization rotator mentioned in Section 3.5, active PVGs can also be achieved by assembling a cell with dual-frequency LC materials [91]. As manifested in Figure 14, after applying a voltage of 50 V (1 kHz), the LC directors are re-oriented to the homeotropic state, where the Bragg grating effect disappears. Switching the frequency of the applied voltage to 50 kHz can turn it back to the Bragg grating state.

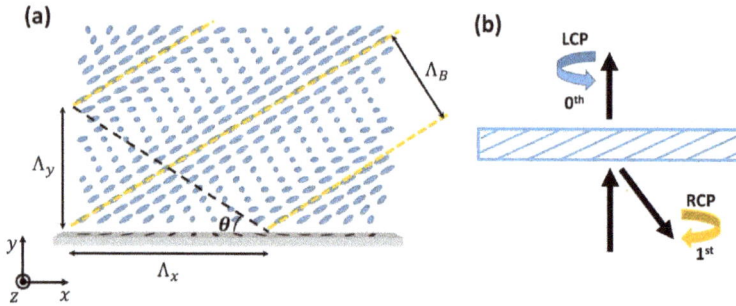

Figure 13. (**a**) Schematic illustration of the LC director distribution in a reflective polarization volume grating (PVG). The yellow dashed lines highlight the Bragg period. (**b**) Polarization selectivity of a right-handed reflective PVG.

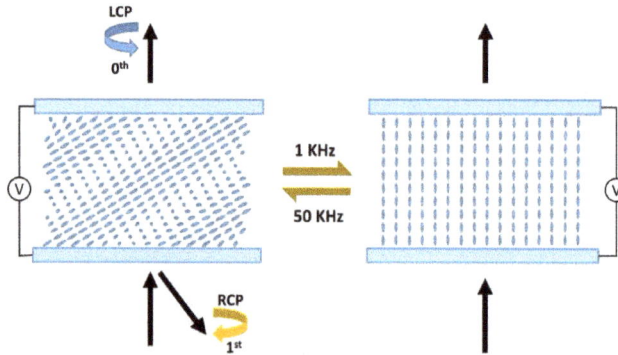

Figure 14. Schematic plot of an electrically-controlled reflective PVG device. Under a 50 V driving voltage, it can be switched between Bragg grating and homeotropic state by applying different AC frequencies.

As mentioned in Section 2, volume gratings possess stronger selectivity in wavelength and angle than thin gratings and this should be considered when targeting specific wavelengths. As a rule of thumb, the central wavelength of a PVG can be precisely designed by the modified Bragg condition:

$$\lambda_B = 2n\Lambda_B \cos\theta \tag{7}$$

where n is the average refractive index of the employed LC materials and θ is the slant angle, defined by $\tan\theta = \Lambda_y/\Lambda_x$. Typically, the bandwidth ranges from 30–50 nm, related to the Δn of the employed LC material. The field of view (angular response) for the central wavelength is around 20°. Fortunately, by applying multi-layer structures, both spectral and angular bandwidths can be dramatically increased while keeping a high diffraction efficiency [92].

4. Future Trends and Challenges

4.1. Fast Response Time for MWIR Beam Steering

As mentioned in Section 3.1, for LC-based beam steering devices (such as OPAs), homogeneous alignment with a positive $\Delta\varepsilon$ LC is preferred. The phase change (δ) of a homogeneous cell is governed by the cell gap (d), wavelength (λ), and bi-refringence (Δn), as:

$$\delta = 2\pi d\Delta n/\lambda \tag{8}$$

On the other hand, the free relaxation time of a homogeneous LC cell is determined by [93]:

$$\tau = \frac{\gamma_1 d^2}{K_{11} \pi^2} \tag{9}$$

where γ_1 is the rotational viscosity and K_{11} the elastic coefficient of the employed LC. In the visible region, a 2π phase change can be easily satisfied. However, as the wavelength increases into the MWIR or long-wave infrared (LWIR) region, as Δn decreases and then gradually saturates [94], a relatively thick LC layer is required, which dramatically increases the response time.

To improve response time while maintaining 2π phase modulation, polymer network liquid crystal (PNLC) has been proven to be an effective approach [58,95]. It consists of a LC host, a small amount of monomer (4–10 wt%), and a photo-initiator (<0.5 wt%) as a photo-polymerizable precursor. After the mixture is filled into a homogeneous LC cell and cured with UV light, a cross-linked polymer network with sub-micron domain size is formed. The polymer network provides strong anchoring energy, which helps to shorten the response time significantly. In this case, the free relaxation time can be estimated using Equation 9, except where d denotes the average domain size. Therefore, a smaller domain size is preferred to accelerate the relaxation. However, a major tradeoff is the increased operating voltage. In order to decrease the voltage, we can minimize the cell gap by employing a high Δn LC.

In addition to response time, LC absorption in the IR region, originating from vibrations of molecular bonds and functional groups, needs to be minimized [96]. To address this issue, we develop an improved eutectic mixture UCF-13, which consists of fluorinated or chlorinated terphenyl LC compounds [58,59,97]. Their chemical structures and compositions are listed in Table 1. The fluoro- (F) and chloro- (Cl) substitutions help shift the vibration absorption band to a longer wavelength and offers good UV stability. Additionally, the terphenyl core structure and terminal cyano (CN) group contribute to high Δn. Therefore, mixture UCF-13 offers low absorption loss, high Δn, and excellent UV stability, which make it a good PNLC candidate for achieving fast response times.

Table 1. Chemical structures and compositions of the proposed mixture UCF-13.

Compound No.	Chemical Structure	Weight (wt%)
1		
		70%
2		
3		20%
4		10%

The wavelength dependent bi-refringence of UCF-13 was measured and the results are plotted in Figure 15a. According to the fitting results by the single-band bi-refringence dispersion equation [94], UCF-13 exhibits a relatively high Δn (0.249) in the MWIR and LWIR regions. Figure 15b depicts the transmittance of UCF-13 in a LC cell with d = 30 μm. In the MWIR region (3–5 μm), although a strong absorption peak occurs at λ = 4.48 μm (due to the CN vibration), its bandwidth is narrow and the baseline transmittance in the off-resonance region (3.8–5.1 μm) is over 96%.

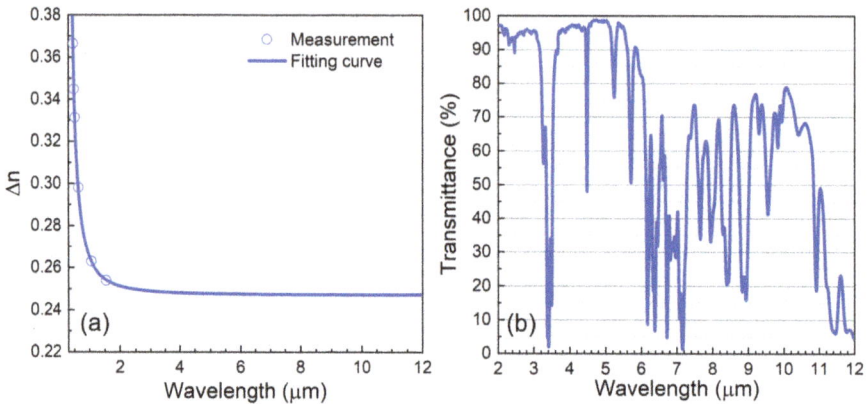

Figure 15. (**a**) Birefringence dispersion of UCF-13 at room temperature: Dots are the measured data and the solid line is the fitting curve using the single-band bi-refringence dispersion equation. (**b**) Measured transmittance spectrum of UCF-13 with a cell gap of about 30 μm.

Figure 16 shows the measured voltage-dependent phase change of our transmissive PNLC sample at λ = 4 μm and four specified temperatures. Although the reflective mode helps the reduce voltage by doubling the optical path length, the transmissive mode is usually preferred, due to its simpler optical system. Besides, the maximum voltage provided by a reflective driving backplane (such as LCoS) is only approximately 24 V [98], which is inadequate for our PNLC. In order to lower the operating voltage for transmissive PNLC, while keeping the response time at millisecond level, we intentionally increase the domain size by controlling the monomer (RM257) concentration at 4.1 wt% and curing temperature at 35 °C. As shown in Figure 16, at T = 25 °C, the voltage for achieving 2π phase change ($V_{2\pi}$) is above 150 V_{rms}. If we increase the temperature to 40 °C or 60 °C, then $V_{2\pi}$ drops to 140 V_{rms} and 150 V_{rms}, respectively, which is achievable by driving the backplane for transmissive devices. As the temperature increases, Δn, $\Delta \varepsilon$, and K_{11} decrease, but at different rates. Thus, 40°C seems to be a good compromise for our PNLC, from a low operation voltage viewpoint.

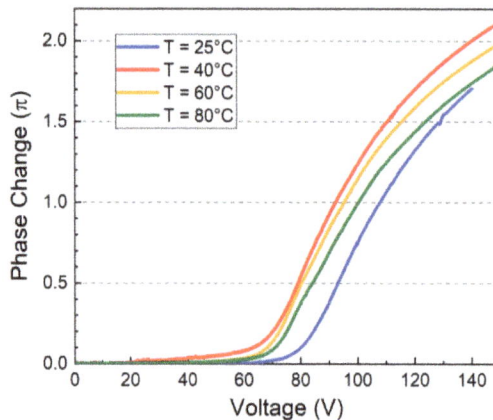

Figure 16. Voltage-dependent phase change of our transmissive polymer network LC (PNLC) device at λ = 4 μm and cell gap 27.5 μm.

To measure relaxation time, we instantaneously removed the applied voltage $V_{2\pi}$ and recorded the transient phase change. The measured relaxation time was calculated between 90% and 10%

phase change. Results for different operating temperatures are plotted in Figure 17a. At $T = 40°C$, the relaxation time was $\tau = 7.3$ ms. If we raised the temperature to 60 °C, then the response time was reduced to $\tau = 1.1$ ms, which is valuable for laser beam steering devices. The fast response time at elevated temperatures was due to decreased γ_1/K_{11} and double relaxation [99]. The solid line shown in Figure 17a is the calculated relaxation time of our PNLC device, according to the following equation [100]:

$$\frac{\gamma_1}{K_{11}} = A \cdot \frac{\exp(E_a/k_B T)}{(1 - T/T_c)^\beta} \tag{10}$$

where A is a proportionality constant, k_B is the Boltzmann constant, E_a is the activation energy, β is the material constant, and T_c is the clearing temperature. At an operating temperature above 60°C, the calculated results fit the experimental data well. At low temperatures, the measured relaxation time was slower than the calculated values. This discrepancy originates from double relaxation. For a polymer-stabilized system, two relaxation processes could take place simultaneously: LC director relaxation and the electrostriction effect of the polymer network. Thus, the relaxation time cannot be completely described by a single exponential term but, instead, by two exponential terms, as:

$$\delta(t) = A \times e^{-2t/\tau_1} + B \times e^{-2t/\tau_2}, \tag{11}$$

where the first and second terms represent the fast and slow relaxation processes, respectively, an [A, B] and [τ_1, τ_2] are the corresponding weights and free relaxation time constants, respectively. The fitting results are shown in Figure 17b. The degree of single relaxation can be quantified as the ratio $A/(A + B)$. If the ratio reaches one (i.e., $B = 0$), we have single exponential decay, because the second term vanishes. For a stronger double relaxation case, B becomes larger, so that $A/(A + B)$ is decreased. In our fitting, the ratio was 90% and the time constants were $\tau_1 = 1.6$ ms and $\tau_2 = 52.4$ ms at $T = 40$ °C. As the temperature increased to 60 °C, this ratio increased to 93% and the time constants were reduced to $\tau_1 = 0.6$ ms and $\tau_2 = 18.0$ ms, indicating a weaker double relaxation. Therefore, an elevated operating temperature helps to suppress the double relaxation and improve the response time of the PNLC device. To achieve a sub-millisecond response time, we could increase the monomer concentration to approximately 5 wt%, but the trade-off is an increased voltage [59].

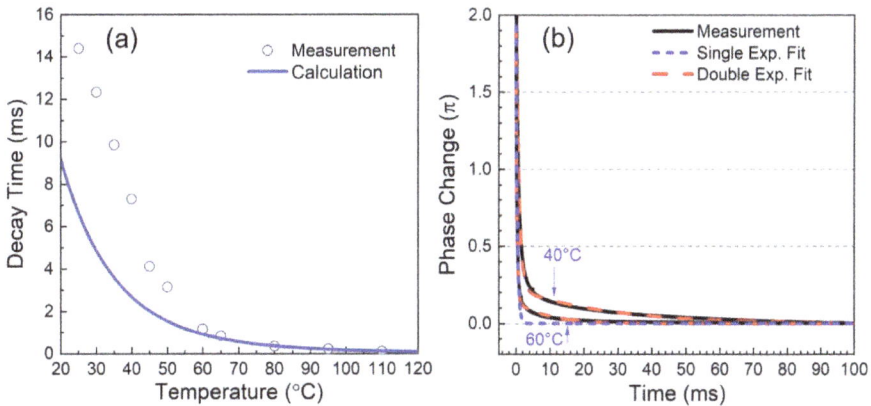

Figure 17. (**a**) Temperature-dependent relaxation time for 2π phase change of our PNLC device. Dots represent the measured data and the solid line is the calculated result using Equation 10. (**b**) Transient relaxation process of the PNLC sample at $T = 40$ °C and 60 °C, fitted with single (dashed blue line) and double (dashed red line) relaxation equations.

Elevated temperature operation not only improves response time, but also reduces hysteresis [99]. Hysteresis is a common phenomenon in polymer-stabilized liquid crystals. It affects grayscale control accuracy and should be minimized. Figure 18 plots the forward and backward voltage-dependent phase changes of the PNLC device at 40 °C and 60 °C, respectively. The hysteresis is calculated using following Equation (12):

$$\Delta h = \frac{V_{\pi,F} - V_{\pi,B}}{V_{2\pi}},\qquad(12)$$

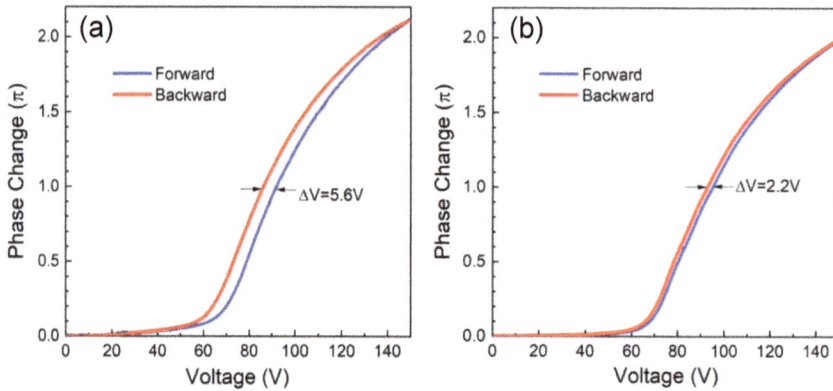

Figure 18. Forward and backward voltage-dependent phase changes of PNLC device at (**a**) $T = 40\,°\text{C}$ and (**b**) $T = 60\,°\text{C}$. In both cases, $\lambda = 4\ \mu\text{m}$ and cell gap 27.5 μm.

In Equation (12), $V_{\pi,F}$ ($V_{\pi,B}$) is the voltage at a π phase change for the forward (backward) scan. According to Figure 18, when the operating temperature increases from 40 °C to 60 °C, the hysteresis Δh decreases from 3.73% to 1.47%.

Besides PNLC, a well-designed scaffold by two-photon polymerization is another promising method to shorten the response time [101]. By artificially creating partition alignment layers in a cell, extra anchoring surfaces can be formed [101,102]. Figure 19 shows a two-layer device. By dividing the LC cell into two sub-layers, the response time should be 4x faster while keeping the same total phase change. Further separating the cell into N LC layers can theoretically obtain an N^2 times faster response time. While the response time improves from the addition of more anchoring surfaces in a cell, the trade-off is increased voltage due to voltage shielding by the polymer partition layers. However, in comparison with PNLC, the major advantage of using a designed scaffold is that one can engineer the response time and driving voltage in a controllable way, as the working voltage increases linearly with the number of polymer anchoring layers. Experimentally, the response time improvements of two-layer and three-layer phase modulators, in reference to a single-layer phase modulator, were 4x and 7x (ideally, it should be 9x) faster, respectively. Some other advantages of multi-layer phase modulators, compared to PNLC, are reduced scattering and diminished hysteresis. Nevertheless, at present, the fabrication yield remains low.

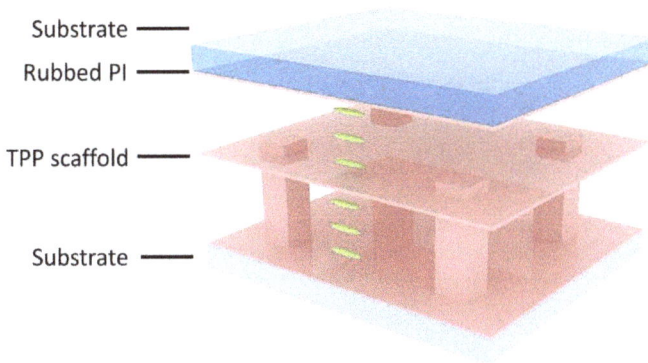

Figure 19. Configuration of a two-layer LC device. The additional polymer anchoring layer with pillar supports can be generated by two-photon polymerization. PI, (polyimide); TPP, (two-photon polymerization).

4.2. Device Hybridization for Continuous, Large-Angle Beam Steering

Realizing continuous, large-angle beam steering devices with high efficiency has been a great challenge for decades. In Section 3, we introduced several devices but, still, none of them can fulfil these performance goals. For example, LC OPAs can achieve about 60% efficiency within a ±5° steering range with quasi-continuous steering angles, and LC-waveguide scanners can continuously scan within 50° (horizontal) by 15° (vertical) with more than 50% efficiency. To achieve a wider field of view, device hybridization is a potential approach.

A straightforward way is to cascade a continuous, tunable device with some discrete high-efficiency devices. For instance, combining an OPA with several PBDs and PVGs is a common approach to fulfil these requirements. Other methods have also been demonstrated, such as combining LC-waveguide scanners with PBDs [56]. Such a hybrid device will, undoubtedly, increase the driving complexity and device bulkiness. However, depending on the applications, these methods are still worth pursuing, because these non-mechanical beam steering devices are highly reliable with long lifetimes and low power consumption.

Another approach is to integrate some small-size electro-mechanical parts, such as MEMS mirrors, with some large-angle beam steerers, such as volume gratings. The utilization of small-size mechanical parts will improve the response time and reliability of the device, compared to the traditional, heavyweight mechanical beam steerers. It is worth mentioning that there have been other interesting hybridizations with mechanical parts, fully benefiting from the elasticity of LC polymers. As mentioned in Section 3.6, PVGs can achieve large diffraction angles with high efficiency. By replacing the substrate of PVGs from glass to PDMS, a stretchable PVG film has been developed [41,103]. Figure 20 depicts the working principle of the PVG film, where the green region depicts the effective grating area. When the PVG film is stretched mechanically along the gray arrow in Figure 20, the grating periodicity will continuously increase, and the corresponding diffraction angle will change accordingly. Releasing the strain will return the film to the un-strained state. However, the tuning range of this device is limited by the bandwidth of the PVGs. Although the diffraction angle of a PVG can reach 70°, the actual steering range is not as large. For example, if the diffraction angle (at a given wavelength) is designed to be 55°, then the maximum reachable steering angle by stretching is about 43.5°, which means that the tuning range is 12°. The response time is dependent on the speed of the mechanical motor and the lifetime of the device still needs further verification.

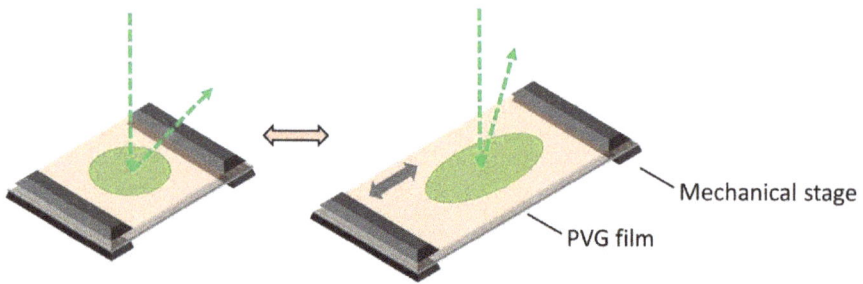

Figure 20. Schematic diagram of a mechanically controlled PVG film for continuous beam steering.

5. Conclusions

We have briefly reviewed the recent advances of LC beam steering devices, including three major operation mechanisms: Blazed gratings (Raman-Nath diffraction), volume gratings (Bragg reflection), and prisms (refraction). In light of these mechanisms, many devices have been proposed and demonstrated. Table 2 summarizes some key performance parameters of the six LC beam steerers mentioned above. For example, OPAs can quasi-continuously steer a laser beam within a ±5° range while keeping diffraction efficiency above 60%. Another promising approach is to use LC-cladding waveguides, which can continuously scan with a fast response time. Improving on the previously-established prism-type LC-cladding waveguide, we proposed a resistive electrode-enabled LC-cladding waveguide in this paper, which can potentially reduce device size. Rising PBDs and PVGs, with nearly 100% diffraction efficiencies, are also strong contenders. Although several devices exhibit high performance in the visible and near IR regions, to extend their useful applications into the MWIR and LWIR regions remains a challenge, with respect to properties such as response time and operation voltage. In this paper, we developed a new transmissive PNLC cell which can achieve a 2π phase change at $\lambda = 4$ µm with a reasonably fast response time; however, its operating voltage (approximately 140 V_{rms}) remains to be improved. Other techniques for addressing these issues are also emerging, such as using a TPP scaffold to create additional partition-anchoring layers in the device. In doing so, both a large phase change and fast response time can be obtained simultaneously. The problem is in the relatively slow fabrication process. On the other hand, to achieve the desired goal of continuous, large-angle beam steering with high efficiency, device hybridization could be a viable approach. We believe that the emergence of novel devices and hybridization methods, as well as the continuous improvement of existing devices, will significantly benefit future beam steering systems.

Table 2. Key parameters of some recently-developed LC beam steering devices.

Devices	Range (^{0})	Efficiency	Continuity	Decay time	λ (µm)	Ref.
OPAs	±5	~60%	Quasi-continuous	a	b	[28,33]
Compound blazed gratings	0–2.3	~40%	Discrete	~ 100's ms	0.532	[48]
	32.1–37.4	~40%	Discrete	~ 100's ms	0.633	[49]
Resistive electrodes	±4.8	NA	Discrete	~ 1's s	0.532	[51]
LC-cladding waveguides	±50	>50%	Continuous	<1 ms	1.55	[56]
	±7	NA	Continuous	NA	4.6	[57]
	±4.9	>90% (c)	Continuous	d	1.55	This work
PBDs	±22	>90%	Discrete	~ 1's ms	1.55	[72]
	±7.6	>90%	Binary	10 ms	4	[74]
PVGs	0–55	>96%	Binary	1.4 ms	0.532	[91]
	43.5–55	>90%	Continuous	e	0.532	[41]

ab. these two parameters depend on each other, as indicated in Equations (8) and (9); *c*: the simulated result is only for reference; *d*: response time depends on the choice of LC materials and a <1 ms response time is achievable; *e*: response time depends on the mechanical motor.

Author Contributions: Methodology, Z.H., F.G. and R.C.; writing—original draft preparation, Z.H., F.G., K.Y. and T.Z.; writing—review and editing, Z.H. and S.-T.W.; supervision, S.-T.W.

Funding: The authors are indebted to the financial support of Air Force Office of Scientific Research (AFOSR) under grant number FA9550-14-1-0279.

Conflicts of Interest: The authors declare no conflict of interest.

References

1. Collis, R.T.H. Lidar. *Appl. Opt.* **1970**, *9*, 1782–1788. [CrossRef] [PubMed]
2. Kattawar, G.W.; Plass, G.N. Time of flight lidar measurements as an ocean probe. *Appl. Opt.* **1972**, *11*, 662–666. [CrossRef]
3. Hair, J.W.; Hostetler, C.A.; Cook, A.L.; Harper, D.B.; Ferrare, R.A.; Mack, T.L.; Welch, W.; Izquierdo, L.R.; Hovis, F.E. Airborne high spectral resolution lidar for profiling aerosol optical properties. *Appl. Opt.* **2008**, *47*, 6734–6752. [CrossRef] [PubMed]
4. Van Kessel, P.F.; Hornbeck, L.J.; Meier, R.E.; Douglass, M.R. A MEMS-based projection display. *Proc. IEEE* **1998**, *86*, 1687–1704. [CrossRef]
5. Lee, Y.H.; Zhan, T.; Wu, S.T. Enhancing the resolution of a near-eye display with a Pancharatnam-Berry phase deflector. *Opt. Lett.* **2017**, *42*, 4732–4735. [CrossRef]
6. Tan, G.; Lee, Y.H.; Zhan, T.; Yang, J.; Liu, S.; Zhao, D.; Wu, S.T. Foveated imaging for near-eye displays. *Opt. Express* **2018**, *26*, 25076–25085. [CrossRef] [PubMed]
7. Betzig, E.; Trautman, J.K. Near-field optics: Microscopy, spectroscopy, and surface modification beyond the diffraction limit. *Science* **1992**, *257*, 189–195. [CrossRef]
8. Neuman, K.C.; Block, S.M. Optical trapping. *Rev. Sci. Instrum.* **2004**, *75*, 2787–2809. [CrossRef]
9. Gattass, R.R.; Mazur, E. Femtosecond laser micromachining in transparent materials. *Nat. Photonics* **2008**, *2*, 219–225. [CrossRef]
10. Lefsky, M.A.; Cohen, W.B.; Parker, G.G.; Harding, D.J. Lidar remote sensing for ecosystem studies. *Bioscience* **2002**, *52*, 19–30. [CrossRef]
11. Næsset, E.; Gobakken, T.; Holmgren, J.; Hyyppä, H.; Hyyppä, J.; Maltamo, M.; Nilsson, M.; Olsson, H.; Persson, Å.; Söderman, U. Laser scanning of forest resources: The nordic experience. *Scand. J. For. Res.* **2004**, *19*, 482–499. [CrossRef]
12. Goodman, J.L. History of space shuttle rendezvous and proximity operations. *J. Spacecr. Rockets* **2006**, *43*, 944–959. [CrossRef]
13. Wehr, A.; Lohr, U. Airborne laser scanning—An introduction and overview. *ISPRS J. Photogramm. Remote Sens.* **1999**, *54*, 68–82. [CrossRef]
14. Duncan, B.D.; Philip, J.B.; Vassili, S. Wide-angle achromatic prism beam steering for infrared countermeasure applications. *Opt. Eng.* **2003**, *42*, 1038–1047.
15. Koh, K.H.; Kobayashi, T.; Lee, C. A 2-D MEMS scanning mirror based on dynamic mixed mode excitation of a piezoelectric PZT thin film S-shaped actuator. *Opt. Express* **2011**, *19*, 13812–13824. [CrossRef] [PubMed]
16. Hofmann, U.; Janes, J.; Quenzer, H.J. High-Q MEMS resonators for laser beam scanning displays. *Micromachines* **2012**, *3*, 509–528. [CrossRef]
17. Chan, T.K.; Megens, M.; Yoo, B.W.; Wyras, J.; Chang-Hasnain, C.J.; Wu, M.C.; Horsley, D.A. Optical beamsteering using an 8×8 MEMS phased array with closed-loop interferometric phase control. *Opt. Express* **2013**, *21*, 2807–2815. [CrossRef]
18. Meyer, R.A. Optical beam steering using a multichannel lithium tantalite crystal. *Appl. Opt.* **1972**, *11*, 613–616. [CrossRef]
19. Nimomiya, Y. Ultrahigh resolving electrooptical prism array light deflectors. *IEEE J. Quantum Electron.* **1973**, *9*, 791–795. [CrossRef]
20. Davis, S.R.; Farca, G.; Rommel, S.D.; Martin, A.W.; Anderson, M.H. Analog, non-mechanical beam-steerer with 80 degrees field of regard. *Proc. SPIE* **2008**, *6971*, 69710G.
21. Römer, G.; Bechtold, P. Electro-optic and acousto-optic laser beam scanners. *Phys. Procedia* **2014**, *56*, 29–39. [CrossRef]
22. Smith, N.R.; Abeysinghe, D.C.; Haus, J.W.; Heikenfeld, J. Agile wide-angle beam steering with electrowetting microprisms. *Opt. Express* **2006**, *14*, 6557–6563. [CrossRef] [PubMed]

23. Liu, C.; Li, L.; Wang, Q.H. Liquid prism for beam tracking and steering. *Opt. Eng.* **2012**, *51*, 114002. [CrossRef]

24. Cheng, J.; Chen, C.L. Adaptive beam tracking and steering via electrowetting-controlled liquid prism. *Appl. Phys. Lett.* **2011**, *99*, 191108. [CrossRef]

25. Kopp, D.; Lehmann, L.; Zappe, H. Optofluidic laser scanner based on a rotating liquid prism. *Appl. Opt.* **2016**, *55*, 2136–2142. [CrossRef] [PubMed]

26. Resler, D.P.; Hobbs, D.S.; Sharp, R.C.; Friedman, L.J.; Dorschner, T.A. High-efficiency liquid-crystal optical phased-array beam steering. *Opt. Lett.* **1996**, *21*, 689–691. [CrossRef] [PubMed]

27. Lindle, J.; Watnik, A.; Cassella, V. Efficient multibeam large-angle nonmechanical laser beam steering from computer-generated holograms rendered on a liquid crystal spatial light modulator. *Appl. Opt.* **2016**, *55*, 4336–4341. [CrossRef]

28. McManamon, P.F.; Bos, P.J.; Escuti, M.J.; Heikenfeld, J.; Serati, S.; Xie, H.; Watson, E.A. A review of phased array steering for narrow-band electrooptical systems. *Proc. IEEE* **2009**, *97*, 1078–1096. [CrossRef]

29. Yang, D.K.; Wu, S.T. *Fundamentals of Liquid Crystal Devices*; Wiley: Hoboken, NJ, USA, 2006.

30. Kato, T.; Mizoshita, N.; Kishimoto, K. Functional liquid-crystalline assemblies: Self-organized soft materials. *Angew. Chem. Int. Ed. Engl.* **2005**, *45*, 38–68. [CrossRef]

31. Woltman, S.J.; Jay, G.D.; Crawford, G.P. Liquid-crystal materials find a new order in biomedical applications. *Nat. Mater.* **2007**, *6*, 929–938. [CrossRef]

32. Bisoyi, H.K.; Li, Q. Light-driven liquid crystalline materials: From photo-induced phase transitions and property modulations to applications. *Chem. Rev.* **2016**, *116*, 15089–15166. [CrossRef] [PubMed]

33. McManamon, P.F.; Dorschner, T.A.; Corkum, D.L.; Friedman, L.J.; Hobbs, D.S.; Holz, M.; Liberman, S.; Nguyen, H.Q.; Pesler, D.P.; Sharp, R.C.; et al. Optical phased array technology. *Proc. IEEE* **1996**, *84*, 268–298. [CrossRef]

34. Zohrabi, M.; Cormack, R.H.; Gopinath, J.T. Wide-angle nonmechanical beam steering using liquid lenses. *Opt. Express* **2016**, *24*, 23798–23809. [CrossRef] [PubMed]

35. He, Z.; Lee, Y.H.; Chanda, D.; Wu, S.T. Adaptive liquid crystal microlens array enabled by two-photon polymerization. *Opt. Express* **2018**, *26*, 21184–21193. [CrossRef] [PubMed]

36. He, Z.; Lee, Y.H.; Chen, R.; Chanda, D.; Wu, S.T. Switchable Pancharatnam-Berry microlens array with nano-imprinted liquid crystal alignment. *Opt. Lett.* **2018**, *43*, 5062–5065. [CrossRef] [PubMed]

37. Nocentini, S.; Martella, D.; Wiersma, D.S.; Parmeggiani, C. Beam steering by liquid crystal elastomer fibres. *Soft Matter* **2017**, *13*, 8590–8596. [CrossRef] [PubMed]

38. Li, C.C.; Chen, C.W.; Yu, C.K.; Jau, H.C.; Lv, J.A.; Qing, X.; Lin, C.F.; Cheng, C.Y.; Wang, C.Y.; Wei, J.; et al. Arbitrary beam steering enabled by photomechanically bendable cholesteric liquid crystal polymers. *Adv. Opt. Mater.* **2017**, *5*, 1600824. [CrossRef]

39. Moharam, M.G.; Young, L. Criterion for Bragg and Raman-Nath diffraction regimes. *Appl. Opt.* **1978**, *17*, 1757–1759. [CrossRef]

40. Saleh, B.; Teich, M. *Fundamentals of Photonics*; Wiley: New York, NY, USA, 2007.

41. Yin, K.; Lee, Y.H.; He, Z.; Wu, S.T. Stretchable, flexible, rollable, and adherable polarization volume grating film. *Opt. Express* **2019**, *27*, 5814–5823. [CrossRef]

42. Weng, Y.; Xu, D.; Zhang, Y.; Li, X.; Wu, S.T. Polarization volume grating with high efficiency and large diffraction angle. *Opt. Express* **2016**, *24*, 17746–17759. [CrossRef]

43. Love, G.D.; Major, J.V.; Purvis, A. Liquid-crystal prisms for tip-tilt adaptive optics. *Opt. Lett.* **1994**, *19*, 1170–1172. [CrossRef] [PubMed]

44. Wu, S.T.; Efron, U.; Hsu, T.Y. Near-infrared-to-visible image conversion using a Si liquid-crystal light valve. *Opt. Lett.* **1988**, *13*, 13–15. [CrossRef] [PubMed]

45. Wang, X.; Wang, B.; Bos, P.J.; McManamon, P.F.; Pouch, J.J.; Miranda, F.A.; Anderson, J.E. Modeling and design of an optimized liquid-crystal optical phased array. *J. Appl. Phys.* **2005**, *98*, 073101. [CrossRef]

46. McManamon, P.F. Agile nonmechanical beam steering. *Opt. Photon. News* **2006**, 21–25. [CrossRef]

47. Wang, X.; Wilson, D.; Muller, R.; Maker, P.; Psaltis, D. Liquid-crystal blazed-grating beam deflector. *Appl. Opt.* **2000**, *39*, 6545–6555. [CrossRef] [PubMed]

48. Shang, X.; Tan, J.Y.; Willekens, O.; De Smet, J.; Joshi, P.; Cuypers, D.; Islamaj, E.; Beeckman, J.; Neyts, K.; Vervaeke, M.; et al. Electrically controllable liquid crystal component for efficient light steering. *IEEE Photonics J.* **2015**, *7*, 1–13. [CrossRef]

49. Willekens, O.; Jia, X.; Vervaeke, M.; Shang, X.; Baghdasaryan, T.; Thienpont, H.; De Smet, H.; Neyts, K.; Beeckman, J. Reflective liquid crystal hybrid beam-steerer. *Opt. Express* **2016**, *24*, 21541–21550. [CrossRef]

50. Klaus, W.; Ide, M.; Morokawa, S.; Tsuchiya, M.; Kamiya, T. Angle-independent beam steering using a liquid crystal grating with multi-resistive electrodes. *Opt. Commun.* **1997**, *138*, 151–157. [CrossRef]

51. Shang, X.; Trinidad, A.M.; Joshi, P.; De Smet, J.; Cuypers, D.; De Smet, H. Tunable optical beam deflection via liquid crystal gradient refractive index generated by highly resistive polymer film. *IEEE Photonics J.* **2016**, *8*, 1–11. [CrossRef]

52. Beeckman, J.; Nys, I.; Willekens, O.; Neyts, K. Optimization of liquid crystal devices based on weakly conductive layers for lensing and beam steering. *J. Appl. Phys.* **2017**, *121*, 023106. [CrossRef]

53. Sun, J.; Xu, S.; Ren, H.; Wu, S.T. Reconfigurable fabrication of scattering-free polymer network liquid crystal prism/grating/lens. *Appl. Phys. Lett.* **2013**, *102*, 161106. [CrossRef]

54. Ren, H.; Xu, S.; Wu, S.T. Gradient polymer network liquid crystal with a large refractive index change. *Opt. Express* **2012**, *20*, 26464–26472. [CrossRef] [PubMed]

55. Davis, S.R.; Farca, G.; Rommel, S.D.; Johnson, S.; Anderson, M.H. Liquid crystal waveguides: New devices enabled by > 1000 waves of optical phase control. *Proc. SPIE* **2010**, *7618*, 76180E.

56. Davis, S.R.; Rommel, S.D.; Johnson, S.; Anderson, M.H.; Anthony, W.Y. Liquid crystal clad waveguide laser scanner and waveguide amplifier for LADAR and sensing applications. *Proc. SPIE* **2015**, *9365*, 93650N.

57. Frantz, J.A.; Myers, J.D.; Bekele, R.Y.; Spillmann, C.M.; Naciri, J.; Kolacz, J.; Gotjen, H.G.; Nguyen, V.Q.; McClain, C.C.; Shaw, L.B.; et al. Chip-based nonmechanical beam steerer in the midwave infrared. *J. Opt. Soc. Am. B* **2018**, *35*, C29–C37. [CrossRef]

58. Peng, F.; Chen, H.; Tripathi, S.; Twieg, R.J.; Wu, S.T. Fast-response infrared phase modulator based on polymer network liquid crystal. *Opt. Mater. Express* **2015**, *5*, 265–273. [CrossRef]

59. Gou, F.; Chen, R.; Hu, M.; Li, J.; Li, J.; An, Z.; Wu, S.T. Submillisecond-response polymer network liquid crystals for mid-infrared applications. *Opt. Mater. Express* **2018**, *26*, 29735–29743. [CrossRef]

60. Tervo, J.; Turunen, J. Paraxial-domain diffractive elements with 100% efficiency based on polarization gratings. *Opt. Lett.* **2000**, *25*, 785–786. [CrossRef]

61. Nikolova, L.; Todorov, T. Diffraction efficiency and selectivity of polarization holographic recording. *Optica Acta* **1984**, *31*, 579–588. [CrossRef]

62. Pancharatnam, S. Generalized theory of interference, and its applications Part I: Coherent pencils. *Proc. Indian Acad. Sci. A* **1956**, *44*, 247–262. [CrossRef]

63. Berry, M. Quantal phase factors accompanying adiabatic changes. *Proc. R. Soc. London Ser. A* **1984**, *392*, 45–57. [CrossRef]

64. Oh, C.; Escuti, M.J. Numerical analysis of polarization gratings using the finite-difference time-domain method. *Phys. Rev. A* **2007**, *76*, 043815. [CrossRef]

65. Lee, Y.H.; Tan, G.; Zhan, T.; Weng, Y.; Liu, G.; Gou, F.; Peng, F.; Tabiryan, N.V.; Gauza, S.; Wu, S.T. Recent progress in Pancharatnam-Berry phase optical elements and the applications for virtual/augmented realities. *Opt. Data Process. Storage* **2017**, *3*, 79–88. [CrossRef]

66. Zhan, T.; Lee, Y.H.; Tan, G.; Xiong, J.; Yin, K.; Gou, F.; Zou, J.; Zhang, N.; Zhao, D.; Yang, J.; et al. Pancharatnam-Berry optical elements for head-up and near-eye displays. *J. Opt. Soc. Am. B* **2019**, *36*, D52–D65. [CrossRef]

67. Sarkissian, H.; Park, B.; Tabirian, N.; Zeldovich, B. Periodically aligned liquid crystal: potential application for projection displays. *Mol. Cryst. Liq. Cryst.* **2006**, *451*, 1–19. [CrossRef]

68. Kim, J.; Li, Y.; Miskiewicz, M.N.; Oh, C.; Kudenov, M.W.; Escuti, M.J. Fabrication of ideal geometric-phase holograms with arbitrary wavefronts. *Optica* **2015**, *2*, 958–964. [CrossRef]

69. Zhan, T.; Xiong, J.; Lee, Y.H.; Chen, R.; Wu, S.T. Fabrication of Pancharatnam-Berry phase optical elements with highly stable polarization holography. *Opt. Express* **2019**, *27*, 2632–2642. [CrossRef]

70. Ichimura, K. Photoalignment of liquid-crystal systems. *Chem. Rev.* **2000**, *100*, 1847–1874. [CrossRef] [PubMed]

71. Nersisyan, S.R.; Tabiryan, N.V.; Steeves, D.M.; Kimball, B.R. The promise of diffractive waveplates. *Opt. Photonics News* **2010**, *21*, 40–45. [CrossRef]

72. Kim, J.; Oh, C.; Serati, S.; Escuti, M.J. Wide-angle, nonmechanical beam steering with high throughput utilizing polarization gratings. *Appl. Opt.* **2011**, *50*, 2636–2639. [CrossRef]

73. Chen, H.; Weng, Y.; Xu, D.; Tabiryan, N.V.; Wu, S.T. Beam steering for virtual/augmented reality displays with a cycloidal diffractive waveplate. *Opt. Express* **2016**, *24*, 7287–7298. [CrossRef]

74. Gou, F.; Peng, F.; Ru, Q.; Lee, Y.H.; Chen, H.; He, Z.; Zhan, T.; Vodopyanov, K.L.; Wu, S.T. Mid-wave infrared beam steering based on high-efficiency liquid crystal diffractive waveplates. *Opt. Express* **2017**, *25*, 22404–22410. [CrossRef] [PubMed]

75. Oh, C.; Escuti, M.J. Achromatic diffraction from polarization gratings with high efficiency. *Opt. Lett.* **2008**, *33*, 2287–2289. [CrossRef]

76. Honma, M.; Nose, T. Temperature-independent achromatic liquid-crystal grating with spatially distributed twisted-nematic orientation. *Appl. Phys. Express* **2012**, *5*, 062501. [CrossRef]

77. He, Z.; Tan, G.; Chanda, D.; Wu, S.T. Novel liquid crystal photonic devices enabled by two-photon polymerization. *Opt. Express* **2019**, *27*, 11472–11491. [CrossRef] [PubMed]

78. Crawford, G.P.; Eakin, J.N.; Radcliffe, M.D.; Callan-Jones, A.; Pelcovits, R.A. Liquid-crystal diffraction gratings using polarization holography alignment techniques. *J. Appl. Phys.* **2005**, *98*, 123102. [CrossRef]

79. Provenzano, C.; Pagliusi, P.; Cipparrone, G. Electrically tunable two-dimensional liquid crystals gratings induced by polarization holography. *Opt. Express* **2007**, *15*, 5872–5878. [CrossRef] [PubMed]

80. Nys, I.; Beeckman, J.; Neyts, K. Switchable 3d liquid crystal grating generated by periodic photo-alignment on both substrates. *Soft Matter* **2015**, *11*, 7802–7808. [CrossRef] [PubMed]

81. Nys, I.; Nersesyan, V.; Beeckman, J.; Neyts, K. Complex liquid crystal superstructures induced by periodic photo-alignment at top and bottom substrates. *Soft Matter* **2018**, *14*, 6892–6902. [CrossRef]

82. Shi, L.; McManamon, P.F.; Bos, P.J. Liquid crystal optical phase plate with a variable in-plane gradient. *J. Appl. Phys.* **2008**, *104*, 033109. [CrossRef]

83. Sutherland, R.L.; Natarajan, L.V.; Tondiglia, V.P.; Bunning, T.J. Bragg gratings in an acrylate polymer consisting of periodic polymer-dispersed liquid-crystal planes. *Chem. Mater.* **1993**, *5*, 1533–1538. [CrossRef]

84. Liu, Y.J.; Sun, X.W. Holographic polymer-dispersed liquid crystals materials, formation, and applications. *Adv. Optoelectron.* **2008**, *2008*, 684349. [CrossRef]

85. Lee, Y.H.; He, Z.; Wu, S.T. Optical properties of reflective liquid crystal polarization volume gratings. *J. Opt. Soc. Am. B* **2019**, *36*, D9–D12. [CrossRef]

86. Kobashi, J.; Mohri, Y.; Yoshida, H.; Ozaki, M. Circularly-polarized, large-angle reflective deflectors based on periodically patterned cholesteric liquid crystals. *Opt. Data Process. Storage* **2017**, *3*, 61–66. [CrossRef]

87. Lee, Y.H.; Yin, K.; Wu, S.T. Reflective polarization volume gratings for high efficiency waveguide-coupling augmented reality displays. *Opt. Express* **2017**, *25*, 27008–27014. [CrossRef] [PubMed]

88. Xiang, X.; Kim, J.; Komanduri, R.; Escuti, M.J. Nanoscale liquid crystal polymer Bragg polarization gratings. *Opt. Express* **2017**, *25*, 19298–19308. [CrossRef]

89. Gao, K.; McGinty, C.; Payson, H.; Berry, S.; Vornehm, J.; Finnemeyer, V.; Roberts, B.; Bos, P. High-efficiency large-angle Pancharatnam phase deflector based on dual-twist design. *Opt. Express* **2017**, *25*, 6283–6293. [CrossRef]

90. Sakhno, O.; Gritsai, Y.; Sahm, H.; Stumpe, J. Fabrication and performance of efficient thin circular polarization gratings with Bragg properties using bulk photo-alignment of a liquid crystalline polymer. *Appl. Phys. B* **2018**, *124*, 52. [CrossRef]

91. Chen, R.; Lee, Y.H.; Zhan, T.; Yin, K.; An, Z.; Wu, S.T. Multistimuli-responsive self-organized liquid crystal bragg gratings. *Adv. Opt. Mater.* **2019**, *7*, 1900101. [CrossRef]

92. Xiang, X.; Kim, J.; Escuti, M.J. Bragg polarization gratings for wide angular bandwidth and high efficiency at steep deflection angles. *Sci. Rep.* **2018**, *8*, 7202. [CrossRef]

93. Wu, S.T. Design of a liquid crystal based tunable electro-optic filter. *Appl. Opt.* **1989**, *28*, 48–52. [CrossRef] [PubMed]

94. Wu, S.T. Birefringence dispersions of liquid crystals. *Phys. Rev. A* **1986**, *33*, 1270–1274. [CrossRef]

95. Sun, J.; Wu, S.T. Recent advances in polymer network liquid crystal spatial light modulators. *J. Polym. Sci.* **2014**, *52*, 183–192. [CrossRef]

96. Wu, S.T. Absorption measurements of liquid crystals in the ultraviolet, visible, and infrared. *J. Appl. Phys.* **1998**, *84*, 4462–4465. [CrossRef]

97. Schadt, M. Liquid crystal materials and liquid crystal displays. *Annu. Rev. Mater. Sci.* **1997**, *27*, 305–379. [CrossRef]

98. Serati, S.A.; Xia, X.; Mughal, O.; Linnenberger, A. High-resolution phase-only spatial light modulators with submillisecond response. *Proc. SPIE* **2003**, *5106*, 138–145.

99. Kneppe, H.; Schneider, F.; Sharma, N.K. Rotational viscosity γ_1 of nematic liquid crystals. *J. Chem. Phys.* **1982**, *77*, 3203–3208. [CrossRef]

100. Kikuchi, H.; Nishiwaki, J.; Kajiyama, T. Mechanism of electro-optical switching hysteresis for (polymer/liquid crystal) composite films. *Polym. J.* **1995**, *27*, 1246–1256. [CrossRef]

101. Lee, Y.H.; Franklin, D.; Gou, F.; Liu, G.; Peng, F.; Chanda, D.; Wu, S.T. Two-photon polymerization enabled multi-layer liquid crystal phase modulator. *Sci. Rep.* **2017**, *7*, 16260. [CrossRef]

102. He, Z.; Lee, Y.H.; Gou, F.; Franklin, D.; Chanda, D.; Wu, S.T. Polarization-independent phase modulators enabled by two-photon polymerization. *Opt. Express* **2017**, *25*, 33688–33694. [CrossRef]

103. Yin, K.; Lee, Y.H.; He, Z.; Wu, S.T. Stretchable, flexible, and adherable polarization volume grating film for waveguide-based augmented reality displays. *J. Soc. Inf. Disp.* **2019**, *27*, 232–237. [CrossRef]

crystals

MDPI

Review

Recent Advances in Adaptive Liquid Crystal Lenses

José Francisco Algorri [1,*], Dimitrios C. Zografopoulos [2], Virginia Urruchi [1] and José Manuel Sánchez-Pena [1]

1   GDAF-UC3M, Displays and Photonics Applications Group, Department of Electronic Technology, Carlos III University of Madrid, Leganés, 28911 Madrid, Spain; vurruchi@ing.uc3m.es (V.U.); jmpena@ing.uc3m.es (J.M.S.-P.)
2   Consiglio Nazionale delle Ricerche, Istituto per la Microelettronica e Microsistemi, 00133 Rome, Italy; dimitrios.zografopoulos@artov.imm.cnr.it
*   Correspondence: jalgorri@ing.uc3m.es; Tel.: +34-916-245-964

Received: 9 May 2019; Accepted: 23 May 2019; Published: 25 May 2019

Abstract: An adaptive-focus lens is a device that is capable of tuning its focal length by means of an external stimulus. Numerous techniques for the demonstration of such devices have been reported thus far. Moving beyond traditional solutions, several new approaches have been proposed in recent years based on the use of liquid crystals, which can have a great impact in emerging applications. This work focuses on the recent advances in liquid crystal lenses with diameters larger than 1 mm. Recent demonstrations and their performance characteristics are reviewed, discussing the advantages and disadvantages of the reported technologies and identifying the challenges and future prospects in the active research field of adaptive-focus liquid crystal (LC) lenses.

Keywords: adaptive-focus lenses; liquid crystal lenses; phase modulation

1. Introduction

Dating back to 1888, the physiologist F. Reinitzer and physicist O. Lehmann described the unusual characteristics of a new state of matter, liquid crystals (LCs) [1]. Since that time and up to more than 70 years later, they remained as purely a scientific curiosity. It was in 1958 when Glen Brown published an article entitled "The Mesomorphic State" in the journal *Chemical Reviews* [2], which reminded the scientific community of the useful LC properties which were revealed years ago by V. K. Freedericksz and V. N. Tsvetkov. Along with the pioneering work of R. Williams and subsequently, of G. H. Heilmeier [3], a revolution was triggered regarding this neglected class of materials. The first screens with LC-based pixels appeared and the investigation of the LC physical and chemical properties was revived. In the next decade, a significant step occurred. Swiss watchmaker BWC (Buttes Watch Company) announced its clock based on LC technology in 1971, which was the first in history. This clock used the Dynamic Scattering principle to drive the display. That same year, M. Schadt and W. Helfrich described a new electro-optic effect of nematic LC, known as a twist effect or TN (twisted nematic) mode [4]. By exploiting this effect, it was possible to fabricate a liquid crystal display (LCD), characterized by low consumption, low values of control voltage, high contrast and relatively fast response speeds [5]. Thanks to the absence of electrochemical degradation of the material, high lifetimes were achieved [6]. In the 1990s, engineers at Hitachi developed active control schemes (active-matrix addressing), thus taking a definitive step toward high-resolution displays [7].

Liquid crystal displays currently monopolize the market and are considered to be the most important LC-based application. Nevertheless, there are many more LC applications which are not related to displays. For example, LC can be used to dynamically tune the response of several components of optical communication networks, such as modulators [8], switches [9], multiplexers [10], filters [11], phase shifters [12], or fiber polarization controllers [13,14]. LC-tunable devices for the aerospace industry [15] or astronomy, performing, for instance, beam steering [16,17] or aberration control [18], have also been developed. The use of LC to control or modify signals at millimeter and sub-millimeter wave frequencies is another research field [19], in which several devices have been proposed, including, among others, tunable antennas [20], phase shifters [21], and filters [22].

Another application sector is the study of the interaction of metamaterials [23,24], nanoparticles [25–27], and carbon nanotubes with LC [28–31]. In particular, the interaction of light with metallic nanoparticles (e.g., gold or silver) presents an interesting phenomenon, known as Localized Surface Plasmon Resonances (LSPR). The LSPR properties can be modified by the presence of an analyte, as in the case of plasmon-based biosensors [32], or by an active medium, such as LCs [33]. When instead of isolated metallic nanoparticles a continuous metal film interacts with light, propagating waves can emerge, known as Surface Plasmon Polaritons (SPP), which provide deep subwavelength confinement. By tuning the SPP properties with an LC material, plasmonic modulators can be designed [34–36]. The LC molecular orientation is very sensitive to changes at the interface between the LC and the surrounding materials. By adding an analyte that allows some kind of binding with the LC or the alignment surface, the LC orientation can be strongly affected by the presence of microparticles such as bacteria, introducing disorder into the area around the binding zone. Hence, LC can make unique optical probes for imaging the molecular ordering and chemical patterns of organic surfaces and sensing the chemical reactions such as enzymatic reactions, DNA hybridization, ligand–receptor bindings, and peptide–lipid interactions at the LC/aqueous interface [37,38]. The same concept is also exploited in other fields, such as security, using LC devices based on different alignments that boost them with particular characteristics of visualization [39].

Furthermore, the dependence of the LC properties on various physical parameters such as deformation, electrical and magnetic fields, pressure and temperature, in addition to the easy integration of LCs in fiber-optics, makes them ideal candidates for the development of distributed sensors [40,41]. For instance, LC are employed in temperature sensing, owing to their high thermo-optic coefficient [42], for instance, in Fabry–Perot systems [43], photonic crystal fibers [44] and chiral nematic polymer networks [45], micrometric structures [46,47], mixtures of LC and metallic nanoparticles in Fabry–Perot [48] or fiber-optic cavities [49].

Finally, LC-tunable optical phase modulation is another application, which has attracted significant attention recently [50], where a large number of techniques and processes have been proposed. Nowadays, it remains a hot research topic, e.g., in ophthalmological applications [51], tunable zooming [52], beam steering [53], correction of aberrations [54], astronomy [55], 3D vision applications [56–59], novel aberration correctors for rectangular apertures [60], micro-axicon arrays [61], multi-optical elements [62], multi-focal [63], high fill-factor [64] and frequency-controlled [65] microlenses, optical vortices [66–69], lensacons, and logarithmic axicons [70]. Among all such applications, adaptive-focus lenses have been the most intensively researched topic during the last 40 years.

The next section describes briefly the state-of-the-art LC-based adaptive lenses. Over the years, numerous concise reviews on the topic of LC lenses have been published. For this reason, this work is focused on recent advances in LC lenses with a diameter larger than 1 mm. As this field is very active, in recent years, several new structures and materials have been proposed, which are reviewed in Section 3. In addition, novel applications have emerged in which LC lenses play a major role. In the last section, we discuss the advantages and disadvantages of the reported technologies and highlight the future prospects and challenges in the technology of adaptive-focus LC lenses.

2. Adaptive-Focus Liquid Crystal Lenses

An adaptive-focus lens is a device capable of tuning its focal length by means of the application of an external stimulus. Conventional lenses rely on two physical parameters in order to modify the impinging wavefront: (a) the difference between the refractive index of the lens material and the surrounding environment and (b) the curvature of their interfaces. For this reason, adaptive-focus lenses are based on devices that change either the refractive index of the lens material or the curvature of its interface. Several techniques for the development of adaptive-focus lenses have been proposed [71]. They can be categorized into three major groups: (a) sliding variable-power rigid lenses, (b) shape-changing fluidic lenses and (c) refractive-index-controlled LC lenses.

In the first group, the optical power is tuned by displacing complementary pieces of identic cubic surface profiles. This technique was independently proposed in the late 1960s by Luis Alvarez [72,73] and Adolph Lohmann [74], and today such lenses are known as Alvarez lenses. Whereas Alvarez' objective was a varifocal ophthalmic lens for presbyopia correction [73], Lohmann searched for a zooming lens system based on lateral rather than longitudinal shifts of lenses [74]. However, refractive Alvarez lenses are very difficult to fabricate due to their cubic surfaces. It was not until 30 years later when a practical device was fabricated by using diffractive plates [75], see Figure 1b. Recently, this technique has been proposed for miniature adaptive lenses [76], and high-speed focusing [77], as shown in Figure 1a. Some advantages are a broad optical power range and the possibility of fast switching.

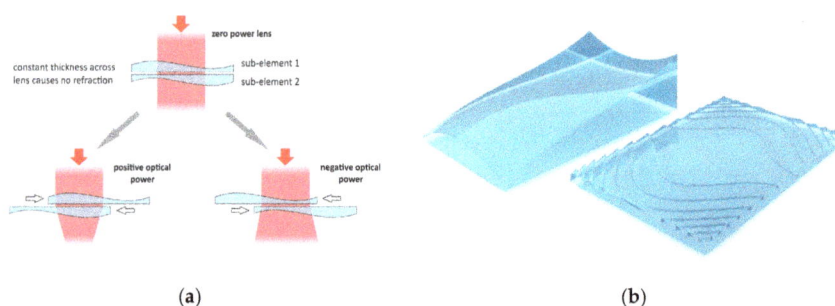

(a)   (b)

Figure 1. Working principle of the Alvarez lens: (**a**) a modulation of the optical power is obtained when conjugate phase plates are shifted with respect to each other in the direction where their phase profile is anti-symmetric. Depending on the displacement of the sub-elements from their zero position, a lens with positive or negative optical power is generated. (**b**) A refractive Alvarez sub-element with a profile function proportional to $x^3/3 + xy^2$ and the corresponding diffractive Alvarez sub-lens, the phase pattern of which is generated by taking the refractive phase structure modulo 2π, are shown. Reprinted with permission from [77]. Copyright 2017 Optical Society of America.

In the case of shape-changing techniques, these can be categorized in the use of elastic membranes, ferrofluidics, soft electroactive actuators, and the electrowetting and dielectrophoretic effect. Depending on the technique, different characteristics can be obtained. For instance, in elastic membrane lenses, a liquid is sandwiched between two substrates of which one is made of an elastic membrane. The lens formation is based on fluidic pressure, by pumping liquid in or out of the lens cavity, which causes changes in the membrane profile accordingly. In order to pump the liquid, a piezoelectric actuator [78] or other mechanical methods [79–82] can be used. For example, in Figure 2a,b the arm of a servomotor deforms the rubber membrane, thus squeezing the liquid contained in the reservoir into the lens chamber. The cavity volume determines the optical power of this kind of lens. The material used to make the membrane is usually polydimethylsiloxane (PDMS). Some current challenges include shape distortion (deflection of the membrane is non-linear), gravitational effect (considerable as the membrane deformation increases), and mechanical actuation. Some improvement can be obtained with piston actuators [83]. Another piston actuator technique is based on ferrofluids [84–86], where the piston is tuned by magnetic fields. Advantages of this approach are large dynamic range and high optical performance. The main problem lies in the precise control of the focal length. One possible solution has been found in the use of soft electroactive actuators [87]. These actuators can be made of different materials, for example, photo-polymer [88–90], hydrogel gel [91,92], conducting polymer artificial muscle [93], ionic polymer–metal composite artificial muscle [94], carbon nanotube artificial muscle [95], dielectric elastomer actuator (DEA) artificial muscle [96,97], and plasticized polyvinyl chloride (PVC) gel [98,99]. The main advantages are the size and easy control (pH, temperature, light, magnetic field, or electric field).

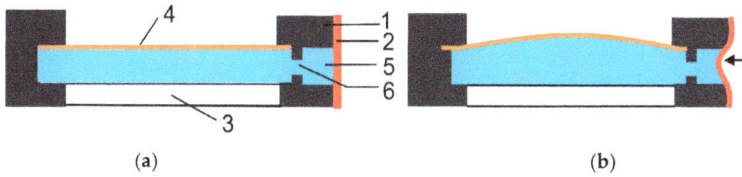

Figure 2. Side view of the liquid lens cell: (**a**) without and (**b**) with a convex lens profile. 1—annular sealing ring, 2—rubber membrane, 3—glass plate, 4—elastic membrane, 5—liquid, and 6—a hole. Reprinted with permission from [80]. Copyright 2006 Optical Society of America.

Other important techniques are electrowetting and dielectric lenses, which are based on liquid–liquid systems. In the latter case, a liquid forms a droplet, which is surrounded by a second liquid of a different type. In electrowetting-based lenses, an electrically conducting liquid drop (usually salt water) is placed in a dielectric-conducting surface, and the macroscopic contact angle can be tuned by changing the charges at the surface. When a control voltage is applied, the droplet tends to spread over a wider area in order to minimize the energy, as schematically shown in Figure 3a,b. Some advantages are the low gravitational effect, stability in case of shocks and vibrations, high optical power, and low response time. In the case of dielectric lenses, two non-conductive liquids with different dielectric constants are used [100,101]. The structure, control, and optical performance are very similar to electrowetting lenses. The main disadvantages of liquid–liquid systems are the operating voltage, the necessary thickness for large aperture lenses (in order to avoid the capillary effect), and possible reflection and scattering due to multiple interfaces.

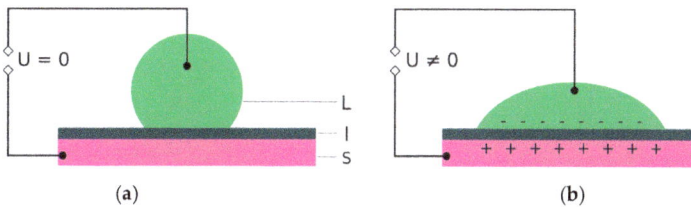

Figure 3. Basic diagram of the electrowetting principle: the change in the solid–electrolyte contact angle due to an applied potential difference between the solid and the electrolyte (**a**) without voltage; (**b**) with applied voltage. (L—Liquid, I—Isolator, S—Substrate, U—Potential). Under CC0 1.0 Universal, public domain dedication.

The last group of adaptive-focus lenses is based on the gradual variation of the refractive index, in principle, without the necessity of any curved interfaces. When light travels through a non-homogeneous medium, the speed of the wavefront decreases in the optically dense regions and accelerates in areas of lower density. Based on this mechanism, lenses without curvature are made by using a material with a spatial gradient of its refractive index; these devices are known as a GRIN (GRadex INdex) lens. In this respect, numerous approaches based on LC-lenses with an electrically controllable focal length have been successfully demonstrated. Many of the topologies proposed for LC lenses are based on generating a gradual voltage across the lens capable of reproducing a parabolic refractive index gradient in the LC layer, thus mimicking the optical behavior of a conventional lens. LC-lenses can substitute their conventional counterparts while being more compact and lightweight. Their true potential, though, lies in applications where a variable focal length is needed, which LC-lenses achieve with low driving voltages and low power consumption.

Among the different LC-based applications, adaptive lenses have been at the forefront for more than 40 years. In the late 1970s, some pioneering research was carried out, giving rise to the first proposals of adaptive lenses, such as the works of Berreman et al. (with patent application in 1977 [102]) and Sato et al. in 1979 [103] (Figure 4a,b). The first adaptive lens was simply based on a curved cavity that was filled with LC. Some problems were the low response time, due to the increased LC layer thickness, and the molecular orientation homogeneity in the employed thick and curved layers. For these reasons, this particular technique was not further developed.

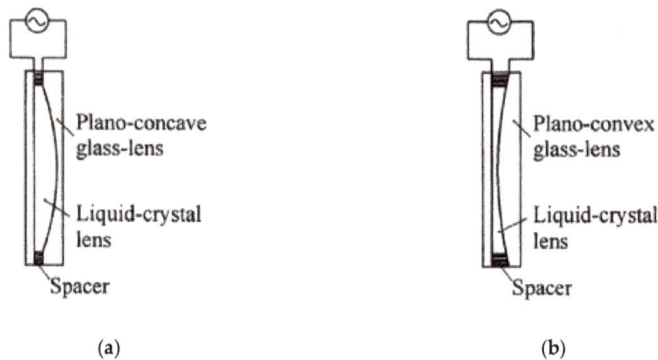

(a)   (b)

Figure 4. First proposals of liquid crystal (LC) lenses. (**a**) Plano-concave lens (**b**) Plano-convex lens. Reprinted with permission from [103]. Copyright 1979 IOP Science.

Some years later, in 1981 [104], the first cylindrical LC lens that was tunable by voltage was demonstrated by a group at Syracuse University, operating as an adaptive optical element. The structure was based on several electrodes that produced the desired voltage gradient. This new concept of lens stimulated other research works [105] and subsequently, spherical lenses [106]. In the late 1980s, new research works demonstrated LC lenses at the micrometric scale [107,108]. Around that time, the first Fresnel lenses were also reported [109]. This configuration reduced the necessary thickness and increased the diameters of the lenses. In the beginning, refractive Fresnel lenses based on surface relief profiles were proposed. Following that, diffractive Fresnel lenses and a mixed version (kinoform lens) were introduced. Also in the 1980s, A.F. Naumov et al. [110] proposed an improved patterned electrode lens to work with large apertures. By using a high-resistivity layer, voltage drop at the edges was avoided [111]. The sheet resistance of the control electrode is a key design parameter. Its value must be in the range between 100 kΩ/sq and a few MΩ/sq for lens diameters in the order of millimeters [112]. This technique has also been proposed for wavefront correctors [54,113], multi-optical devices [114,115] or optical tweezers [116,117]. Intense research on this technique resulted in a commercial device by Flexible Optical B.V. (OKO Tech) [118] (Figure 5).

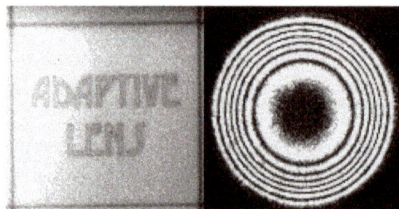

Figure 5. Commercial modal LC lens by Flexible Optical B.V. Reprinted with permission from [118].

In the following years, an exponential increase in research on LC lenses was experienced, which continues to this day, making adaptive-focus lenses one of the most important and researched topics in the field of LCs during the last 40 years. Over these years, several reviews have been published. Some recent reviews are focused on specific topics, such as fast response time LC microlenses [119], LC microlenses for autostereoscopic displays [120], design and fabrication [121] or LC contact lenses for the correction of presbyopia [122]. As a general approach, the review on LC lenses by Lin et al. is by far the most complete and detailed [123]. For this reason, the next section is only focused on recent developments in LC lenses with a diameter larger than 1 mm. Despite the fact that the main structures were established several years ago, there is still room for new proposals.

3. Recent Developments in Liquid Crystal Lenses

Despite the fact that LC lenses were first reported 40 years ago, they still remain an active field of research. In general, the applications of LC lenses are as numerous as those of fixed lenses or GRIN lenses, with the great advantages of volume and weight reduction and tunable focal distance by voltage. For example, LC lenses can help realize auto-focusing and optical zoom systems [124]. They have also been proposed for pico-projection systems, helping to electrically adjust the focusing properties of the projected image without mechanically adjusting the position of a projection lens [125]. In the case of holographic projection systems, LC lenses can help to correct the mismatch of chromatic image size, which is important for full-color holographic projection systems [126]. LC lenses can also be used as a concentrator and a sun tracker in a concentrating photovoltaic (CPV) systems [127] or couplers for fiber optics [128].

Furthermore, it is worth mentioning the use of LC lenses in the field of bio-optics, e.g., as optical tweezers [116,129], or in medical instrumentation in applications such as endoscopy, where LC lenses can be adopted to electrically enlarge the depth-of-field of an endoscopic system [130]. In addition, when it comes to ophthalmic lenses, the lens power of LC lenses is not only electrically tunable but it can also be positive or negative. LC lenses can correct myopia, hyperopia, and presbyopia. In the case of novel augmented reality displays, it can alleviate problems for people with disabilities and resolve the issue of accommodation–convergence [131]. In this regard, contact lenses made from LCs have been proposed [132–135], as well as LC-based commercial glasses, which were developed in 2011 [136] (Figure 6). Although such glasses tackle the problem of patients who require different types of glasses for different activities, the company went bankrupt due to a high rate of return (battery issues and some defective devices). LC lenses can also be considered as a kind of "extra-artificial crystalline lens" to compensate for the degradation of the crystalline lens of aging eyes or eye accommodation [137].

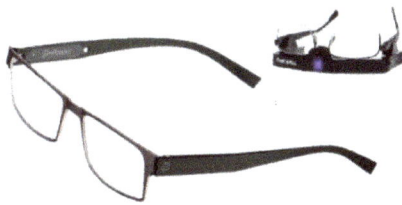

Figure 6. Commercial LC glasses with tunable focus (PixelOptics) [136].

In short, in recent years there have been lots of developments tailored to the needs of different applications. These works can be grouped as a function of the employed technique. In this review, they are divided into the following categories: curved lenses, patterned electrodes, modal control, alignment layer, Fresnel, and polarization-independent lenses.

3.1. Curved Lenses

As previously mentioned, this technique was the first proposal for an LC lens due to the similarity to classic lenses. Despite this, in recent years intense research on applying this concept to contact lenses has been produced. For example, a switchable polymethyl methacrylate (PMMA) LC lens to function as a contact lens was proposed, as shown in Figure 7a,b [134]. The device has an active optical zone of 4 mm and it produces a variable focal power of up to +2.00 diopters, perfect for presbyopia correction. The same group has demonstrated that the use of negative LC and vertical alignment in the same structure provides a continuous change in focal power up to −2.00 [138]. The design considerations for this type of lens are reported in [139]. Other proposals that rely on the use of flexible substrates can be found in [140], using polyethylene terephthalate (PET) as a substrate. Two display substrates with different surface areas are used, allowing for the integration of powering and driving electronics (the purpose is also to work as a contact lens). The authors claim that ring-shaped concentric pixels, which are able to switch from a transparent to an opaque state, can be used to mimic an ocular iris and that it could be used by people suffering from iris deficiencies that involve hypersensitivity to light, such as aniridia, coloboma, or ocular albinism [140].

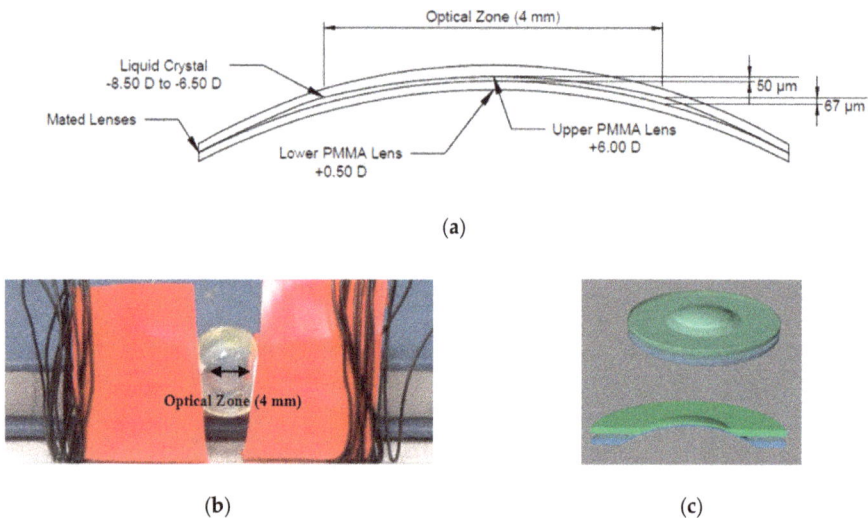

(a)

(b)

(c)

Figure 7. (**a**) Shows a diagram of the design used for the switchable LC lens. The base curve of the lens is designed to fit onto the eye, with a radius of curvature of 7.8 mm, and therefore, appropriate for the average human cornea. The optical power of the lower substrate is +0.50 D, the optical power of the upper substrate is +6.00 D, with the optical power of the LC layer varying between −8.50 D and −6.50 D, depending on the polarization direction of light and the orientation of LC molecules. (**b**) A photograph of the device. (**c**) CAD rendered images of the lens. Reprinted with permission from [134]. Copyright 2014 Optical Society of America.

Another option is to embed dielectric curved elements inside the cell. This technique has been used extensively to produce LC microlenses; however, the main challenge is the necessary thickness of these elements. An array of LC lenses with diameters of 1.2 mm is demonstrated in [141]. The optical power can be switched from +100 to +50 diopters for voltages ranging from 11 to 14 V_{RMS}, respectively.

3.2. Patterned Electrode

This category comprises designs that have patterned Indium tin oxide (ITO) in the structure. This was the first technique proposed to design LC microlenses, and involved exploiting a basic mechanism. A hole in the ITO layer produces a voltage decrease toward the lens center that has a quasi-parabolic profile. The main problem is that for diameters larger than the thickness of the sample, the voltage drops at the edges. Several structures have been recently proposed to counter this problem. In [142,143], a floating ring electrode is embedded in the interface between the dielectric layer and the LC layer. A diameter of 6 mm was reported and the operating voltage range was from 0 to 40 V_{RMS}. For the maximum voltage, an optical power of +8.33 diopters was achieved. A different approach has been proposed in [144], by using internal resistances between electrodes to conduct the voltage towards the center. In [145], a multielectrode configuration is proposed, which is different from previous designs in that only one lithographic step is necessary (Figure 8a–d). All the electrodes are located on the same layer. Although it is a microlens, it is included because the diameter can easily be scaled to larger values. The main problem is that the higher the number of control voltages, the larger the defect area (see Figure 8b) in the active region in the lens area, thus limiting the maximum diameter.

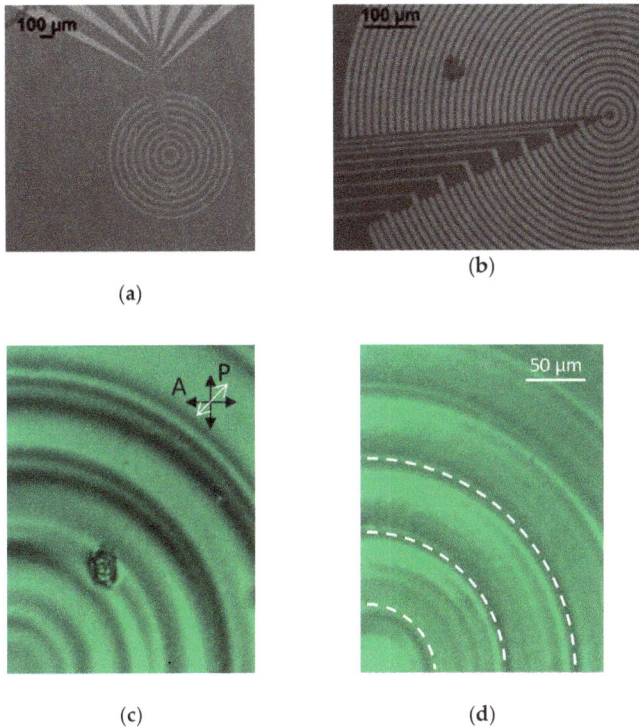

(a)

(b)

(c)

(d)

Figure 8. Reflection microscope image of the ITO electrode pattern for a lens (**a**) without floating electrodes and (**b**) with floating electrodes. Polarization microscope image of the LC devices with PZT (**c**) without floating electrodes and (**d**) with floating electrodes when 5 V is applied to all the electrodes. The rubbing direction is denoted with a white arrow. The position of the contacted electrodes is indicated with a white dashed line. Reprinted with permission from [145]. Copyright 2018 Optical Society of America.

3.3. Modal Control

This technique was also among one of the first ones to be proposed, and the advances during the last 40 years are related to the use of different materials for the high-resistivity layer, among which titanium oxide films [146], thin ITO layers [147] (max. 10 MΩ/sq [148]) or PEDOT [149].

In recent years, novel materials have been proposed, such as those used in [150], where a multi-functional liquid-crystal lens based on a dual-layer electrode design have been demonstrated. The diameter was 1.42 mm and the material for the high-resistivity layer was niobium pentoxide (Nb_2O_5). As with other materials, the optimization of the fabrication process is key in order to obtain stable parameters. Depending on the conditions, a broad range of sheet resistance values are obtained. One of the main problems of modal control, despite being the best technique in terms of simplicity and power consumption, is that the manufacturing of homogenous high-resistivity layers is not trivial to obtain. A high-resistivity layer of ZnO (100 MΩ/sq) is employed in [151], where the thickness of the LC layer is 30 μm, the diameter 2 mm, and the optical power can be tuned from −3.9 to +4.4 diopters by using two different voltages. In [152], a PEDOT layer is used, but the electro-optical features of the LC lens are improved by doping the LC with multi-walled carbon nanotubes. The thickness of the LC layer is 100 μm and the diameter is 2.3 mm. When the external applied voltage is adjusted from 0 to 5 V_{RMS}, the lens power switches from +1.92 to +83.33 diopters. Additionally, the response time is reduced to 0.1 s in comparison with 2.6 s for non-doped cells. Another material recently proposed is an oxide Ag alloy [153]. Moreover, as claimed by the authors, the addition of a floating electrode allows significant optimization of the spatial distribution of the electrical potential (Figure 9a,b). The spherical aberration is reduced considerably, thus obtaining a high-quality lens. The lens has been tested in an 8-megapixel miniature camera of a cell phone [154]. Finally, different modifications for the modal control technique (based on inner and floating electrodes) are analyzed in [155]. It is claimed that it is possible to obtain a parabolic wavefront across the entire clear aperture of the lens. The proposed design has a 4 mm clear aperture, very low RMS spherical aberrations, and a dynamic range of +3.5 diopters.

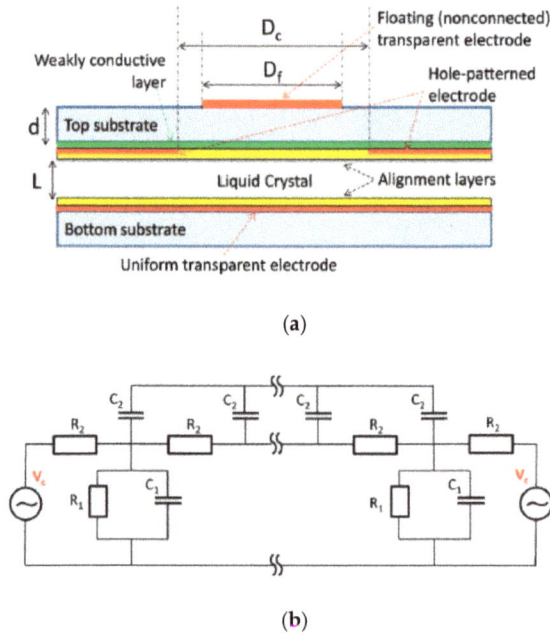

(a)

(b)

Figure 9. (**a**) Schematic presentation of the structure of the proposed modal LC lens, and (**b**) its approximate equivalent electrical circuit. V_c is the control voltage (applied on the hole patterned electrode); R_1 and C_1, are the equivalent electric resistance and capacitance of the LC cell, respectively; R_2, is the equivalent resistance of the high-resistivity layer; C_2, the equivalent capacitance due to the floating electrode. Reprinted with permission from [153]. Copyright 2016 Optical Society of America.

3.4. Aligment Layer

This technique is usually based on the use of polymers which are sensitive to UV radiation as alignment layers [156]. Varying the UV dosage controls the pretilt angle. In recent years, new materials have been investigated, typically, these have been photosensitive polymers with benzaldehyde side groups. In references [157,158], a variable pretilt in the range from 0° to 90° is achieved via the formation of a localized polymer network. Other examples have used a new polymer characterized by the percentage of the three components and its macromolecules, comprising of photosensitive benzaldehyde-containing fragments (B), methacrylate fragments, causing vertical LC alignment (VA), and methacrylate fragments, producing planar LC alignment under photocrosslinking of the material, as investigated in [159]. The main difference with other materials is the two-step treatment: uniform rubbing (defines the azimuthal angle) and gradient nonpolarized UV exposure (to define the pretilt angle). Photocrosslinkable benzaldehyde polymers with fragments including long hydrocarbon substituents in a side chain are used in [160].

A different approach is the use of photoalignment to create Pancharatnam–Berry lenses. In [161], an LC lens with a diameter of 16 mm and an optical power of +27.7 diopters (99% diffraction efficiency) is reported. The thickness of the LC layer is only a few µm so the response time is reduced. As reported in [162], this kind of lens exhibits a fast switching time between two or more focal lengths with a large diopter change and aperture size (Figure 10a–d). In order to fabricate such a lens, a photo-alignment material is coated onto the substrate. The cell is mounted and subject to interference exposure. Another method to produce Pancharatnam–Berry lenses is proposed in [163]. In this case, a ferroelectric liquid crystal (FLC) cell with a continuous alignment structure is realized by a polarization holographic method. Authors report a diffraction efficiency of up to 87% and a response time of 300 µs with a low electric voltage (4 V/µm). The optical power can reach +2.4 diopters for a 13 mm aperture. This type of lens has been proposed to solve the problem of accommodation–convergence in head-mounted display devices [164,165] and to generate spatially separated focuses [166].

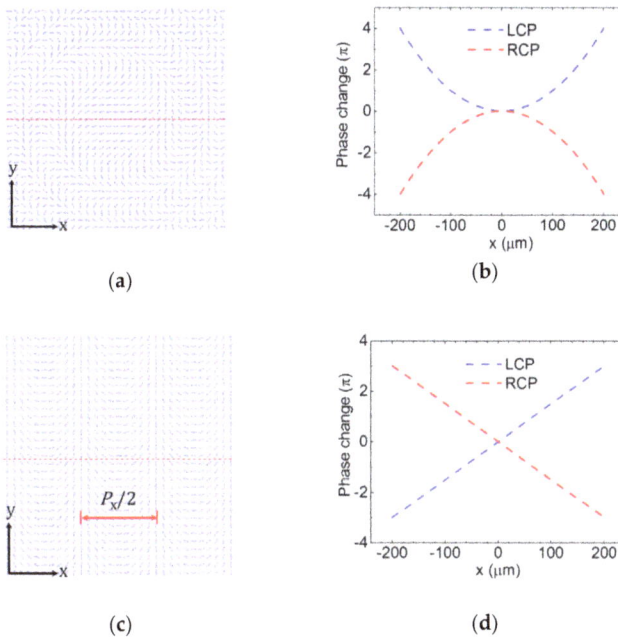

Figure 10. (a) Top view of a Pancharatnam-Berry lens: the direction of the LC optical axis changes radially. (b) The corresponding phase change along the dashed lines in (a). (c) Top view of a Pancharatnam-Berry deflector: the direction of the LC optical axis changes longitudinally. (d) The corresponding phase change along the dashed lines in (c). Reprinted with permission from [162]. Copyright 2017 Optical Society of America.

3.5. Fresnel

Fresnel lenses have been investigated since the beginning of research on LC lenses. It is the most viable option to obtain high optical powers with large aperture diameters. At the beginning, refractive Fresnel lenses based on surface relief profiles were proposed. After that, diffractive Fresnel lenses and a mixed version (kinoform lens) were introduced. In these 40 years, there have been several proposals to obtain this type of structure.

In recent years, some cases include refractive Fresnel lenses [167] and diffractive Fresnel lenses, in which the zones are made by confining ferroelectric liquid crystals (FLCs) in multiple microscopically defined photo-aligned alignment domains [168] (Figure 11a,b); by multi-electrodes [169]; polymer-dispersed liquid crystal [170]; by a 90° twisted-nematic liquid crystal (LC) cell with a photoconductive polymer layer [171]; by using a photoconductive polymer layer and a Sagnac interferometer [172]; by a P6CB alignment layer (orientation direction also controlled along the polarization direction of UV light) and an interferometric setup [173]; by a patterned hybrid photo-aligned nematic dual frequency LC [174]; by Polymer-Stabilized Blue Phase (PSBP) LC zones [175–177] (the lenses being also polarization-independent). In these cases, the thickness is considerably reduced (between 1.5 um to 12 um), and therefore, the switching is fast. The optical power can be very high ranging from +1.5 to +23 diopters. The main disadvantage of these designs is the low diffraction efficiency obtained (around 30–40% for a binary phase lens).

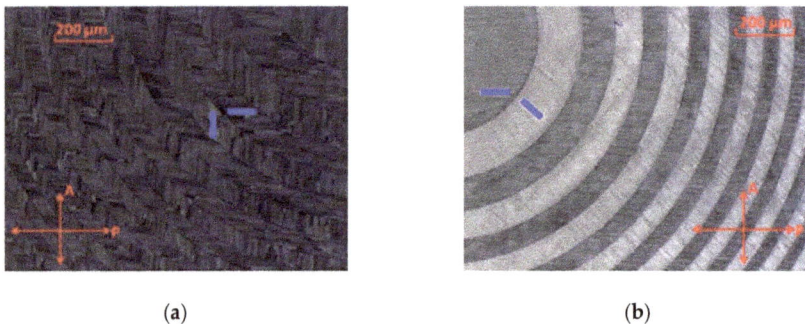

(a)   (b)

Figure 11. Microphotographs of two designed ferroelectric liquid crystal Fresnel zone lens under crossed polarizers. The blue lines indicate the alignment directions. (**a**) ferroelectric liquid crystal Fresnel zone lens (Rin 360 μm) with β 90°. (**b**) ferroelectric liquid crystal Fresnel zone lens (Rin 255 μm) with β 45°. Reprinted with permission from [168]. Copyright 2015 Optical Society of America.

3.6. Polarization-Independent Lenses

Polarization-independent lenses are very interesting from a practical point of view. They do not require external polarizers and usually provide high optical power. As commented in the previous section, PSBP-LC Fresnel lenses are one example of polarization-independent lenses, yet other approaches based on this material can be found in [178], where a matched conventional glass lens is introduced in a Blue Phase Liquid Crystal (BPLC) lens to increase the range of the tunable focal length. The main problem associated with this material is the high voltage required to change the refractive index. In addition to PSBP LC, this effect can be achieved by stacking two LC lenses orthogonally. In recent years, some advanced structures have been proposed. For example, by stacking a number of LC layers, the aperture size of the LC lens can be enlarged without lowering the tunable lens power [179] (Figure 12a–c).

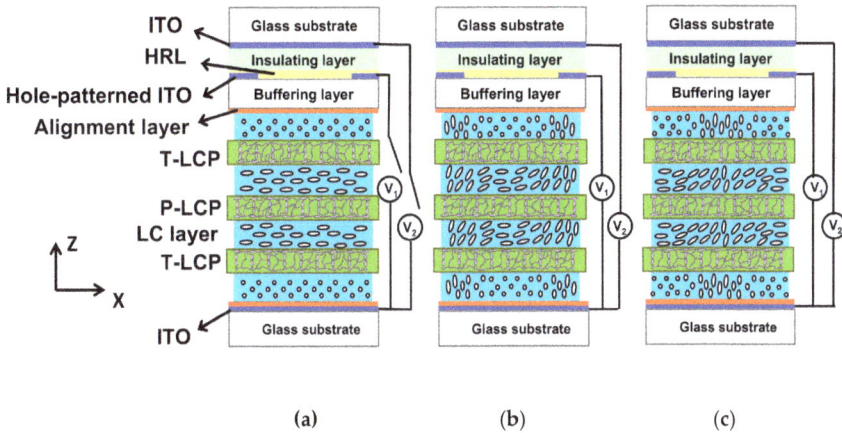

Figure 12. Schematic diagrams of the operation of the multilayered LC lens at (a) voltage-off state, (b) $V_1 > V_2$ and (c) $V_2 > V_1$. © 2015 IEEE. Reprinted, with permission, from [179].

For a large aperture of 10 mm, the reported switching time is ~3 s for imaging objects from 360 cm to 26 cm. The lens power is +3.93 diopters for a driving voltage of 80 V_{RMS}. These are promising values for an LC lens with such a diameter. This design can solve several classic problems of LC lenses, such as the necessity of thick devices for high optical power, the derived response time of thick devices, and the polarization dependency of LC lenses. Another recently demonstrated approach is by using Pancharatnam–Berry lenses with a specific configuration [180]. The structure consists of at least four Pancharatnam–Berry lenses, with specific distances between them. As the authors claim, by compensating the polarization dependency of each Pancharatnam–Berry lens, single-depth high-resolution images can be obtained without preprocessing the incident polarization state.

4. Discussion

Adaptive-focus lenses have attracted a great amount of scientific and technological interest in recent years. Each technique discussed in this short review has its own strengths and weaknesses. For small apertures (less than 5 mm), LC lenses have been demonstrated to be the most suitable option. Some techniques are very easy to fabricate, the operating voltages and power consumption are very low and the long-term stability is very high owing to the lack of mechanically moving parts. The main problem of larger apertures is that, for the same gradient of the refractive index, the higher the aperture the lower the optical power. Another problem is the thickness, which has to be large in order to obtain a large dynamic focus range, and thin to obtain a low response time, thus defining a trade-off in performance. For larger apertures, liquid lenses could be more appropriate, as the optical path length change is high due to the change in the shape of the material. Moreover, liquid lenses have lower response times and are polarization-independent. In this regard, some proposals for LC lenses as multilayer lenses have solved the problem of polarization dependency and switching time. For high optical powers, curved lenses appear to be a good option, as demonstrated by recent proposals that offer flexible solutions for contact lenses. Fresnel lenses can also produce high optical powers, but without the possibility of analog tunability. Many different materials have been proposed in recent years. When the optical power or response times are not the main requirement, modal lenses are a promising option since they feature lower power consumption, low voltage control, and simple fabrication. Some new materials for high-resistivity layers and minor modifications to the structure have demonstrated high-quality lensing and a dynamic focus range (from −3.9 to +4.4 diopters for an aperture of 2 mm). Furthermore, other works have shown the potential of parabolic phase profiles for a 4 mm diameter with very low RMS spherical aberrations and a dynamic range of +3.5 diopters.

Finally, new designs based on the smart control of alignment layers have appeared recently. The Pancharatnam–Berry lenses are worth mentioning as they exhibit a fast switching time between two or more focal lengths, high optical power for large apertures, and very thin layers, in the order of a few micrometers. These properties make them suitable for new applications, such as augmented reality devices.

In conclusion, LC lenses continue to attract the interest of researchers, mainly owing to the large number of applications in which they can be applied. Some new devices, such as head-mounted displays, mobile cameras, or contact lenses have demonstrated the need for novel adaptive lenses. One of the main challenges in the field of LC lenses is achieving the ability to increase the apertures to more than 10 mm while maintaining a high optical power (more than 3 diopters). Depending on the specific application, some parameters such as response time, dispersion, operation voltage, and phase profile must be taken into account. In the following years, novel materials and structures can potentially solve most of these problems, thus making LC lenses one of the most suitable platforms for the design of a novel class of adaptive-focus lenses for emerging applications.

Author Contributions: All authors contributed equally to the manuscript.

Funding: This work was supported by Comunidad de Madrid and FEDER Program under grant S2018/NMT-4326 and the Ministerio de Economía y Competitividad of Spain (TEC2013-47342-C2-2-R).

Conflicts of Interest: The authors declare no conflict of interest.

References

1. Lehmann, O. Über fliessende Krystalle (On flowing crystals). *Z. Phys. Chem.* **1889**, *4*, 462–472.
2. Friedel, M.G. The Mesomorphic states of matter. *Ann. Phys. (Paris)* **1922**, *18*, 162–174.
3. Heilmeier, G.H. Guest-host interactions in nematic liquid crystal. A new electro-optic effect. *Appl. Phys. Lett.* **1968**, *13*, 91. [CrossRef]
4. Schadt, M.; Helfrich, W. Voltage-dependent optical activity of a twisted nematic liquid crystal. *Appl. Phys. Lett.* **1971**, *18*, 127. [CrossRef]
5. Kawamoto, H. The history of liquid-crystal displays. *Proc. IEEE* **2002**, *90*, 460–500. [CrossRef]
6. Libsh, F.R.; Kanichi, J. TFT lifetime in LCD operation. In *1993 SID International Symposium: Digest of Technical Papers*; Society for Information Display: Playa del Rey, CA, USA, 1993.
7. Lueder, E. *Liquid Crystal Displays: Addressing Schemes and Electro-Optical Effects (Wiley Series in Display Technology)*; Wiley: Hoboken, NJ, USA, 2001; ISBN 0471490296.
8. Residori, S.; Bortolozzo, U.; Peigné, A.; Molin, S.; Nouchi, P.; Dolfi, D.; Huignard, J.P. Liquid crystals for optical modulation and sensing applications. In Proceedings of the SPIE 9940, Liquid Crystals XX, 99400N, San Diego, CA, USA, 22 November 2016.
9. Pinzón, P.J.; Pérez, I.; Vázquez, C.; Sánchez Pena, J.M. Reconfigurable 1 × 2 wavelength selective switch using high birefringence nematic liquid crystals. *Appl. Opt.* **2012**, *51*, 5960. [PubMed]
10. Lallana, P.C.; Vázquez, C.; Vinouze, B.; Heggarty, K.; Montero, D.S. Multiplexer and variable optical attenuator based on PDLC for polymer optical fiber Networks. *Mol. Cryst. Liq. Cryst.* **2009**, *502*, 130–142. [CrossRef]
11. Pinzon, P.J.; Vazquez, C.; Perez, I.; Sanchez Pena, J.M. Synthesis of Asymmetric Flat-Top Birefringent Interleaver Based on Digital Filter Design and Genetic Algorithm. *IEEE Photonics J.* **2013**, *5*, 7100113. [CrossRef]
12. Koeberle, M.; Hoefle, M.; Chen, M.; Penirschke, A.; Jakoby, R. Electrically tunable Liquid Crystal phase shifter in antipodal finline technology for reconfigurable W-Band Vivaldi antenna array concepts. *Antennas Propag.* **2011**, 1536–1539.
13. Zografopoulos, D.C.; Kriezis, E.E. Tunable Polarization Properties of Hybrid-Guiding Liquid-Crystal Photonic Crystal Fibers. *J. Light. Technol.* **2009**, *27*, 773–779. [CrossRef]
14. Pitilakis, A.K.; Zografopoulos, D.C.; Kriezis, E.E. In-Line Polarization Controller Based on Liquid-Crystal Photonic Crystal Fibers. *J. Light. Technol.* **2011**, *29*, 2560–2569. [CrossRef]

15. Geday, M.A.; Quintana, X.; Otón, E.; Cerrolaza, B.; Lopez, D.; de Quiro, F.G.; Manolis, I.G.; Short, A. Development of liquid crystal based adaptive optical elements for space applications. In Proceedings of the ICSO 2010 International Conference on Space Optics, Rhodes, Greece, 4–8 October 2010.

16. Oton, E.; Morawiak, P.; Mazur, R.; Quintana, X.; Geday, M.A.; Oton, J.M.; Piecek, W. Diffractive and refractive liquid crystal devices based on multilayer matrices. *J. Light. Technol.* **2019**, *37*, 2086–2093. [CrossRef]

17. Oton, E.; Perez-Fernandez, J.; Lopez-Molina, D.; Quintana, X.; Oton, J.M.; Geday, M.A. Reliability of liquid Crystals in Space Photonics. *IEEE Photonics J.* **2015**, *7*, 1–9. [CrossRef]

18. Hu, L.; Xuan, L.; Cao, Z.; Mu, Q.; Peng, Z.; Liu, Y.; Yao, L.; Yang, C.; Lu, X.; Xia, D.L.M. The state of the art for liquid crystal adaptive optics in astronomical applications. In Proceedings of the 2012 International Conference on Optoelectronics and Microelectronics, Changchun, China, 23–25 August 2012; pp. 428–430.

19. Zografopoulos, D.C.; Ferraro, A.; Beccherelli, R. Liquid-crystal high-frequency microwave technology: materials and characterization. *Adv. Mater. Technol.* **2018**, *4*, 1800447. [CrossRef]

20. Gaebler, A.; Moessinger, A.; Goelden, F.; Manabe, A.; Goebel, M.; Follmann, R.; Koether, D.; Modes, C.; Kipka, A.; Deckelmann, M.; et al. Liquid crystal-reconfigurable antenna concepts for space applications at microwave and millimeter waves. *Int. J. Antennas Propag.* **2009**, *2009*, 1–7. [CrossRef]

21. Muller, S.; Scheele, P.; Weil, C.; Wittek, M.; Hock, C.; Jakoby, R. Tunable passive phase shifter for microwave applications using highly anisotropic liquid crystals. In Proceedings of the IEEE MTT-S International Microwave Symposium Digest (IEEE Cat. No.04CH37535), Fort Worth, TX, USA, 6–11 June 2004.

22. Urruchi, V.; Marcos, C.; Torrecilla, J.; Sánchez-Pena, J.M.; Garbat, K. Note: Tunable notch filter based on liquid crystal technology for microwave applications. *Rev. Sci. Instrum.* **2013**, *84*, 026102. [CrossRef] [PubMed]

23. Zografopoulos, D.C.; Beccherelli, R. Tunable terahertz fishnet metamaterials based on thin nematic liquid crystal layers for fast switching. *Sci. Rep.* **2015**, *5*, 13137. [CrossRef] [PubMed]

24. Isić, G.; Vasić, B.; Zografopoulos, D.C.; Beccherelli, R.; Gajić, R. Electrically tunable critically coupled terahertz metamaterial absorber based on nematic liquid crystals. *Phys. Rev. Appl.* **2015**, *3*, 064007. [CrossRef]

25. Hegmann, T.; Qi, H.; Marx, V.M. Nanoparticles in liquid crystals: Synthesis, self-assembly, defect formation and potential applications. *J. Inorg. Organomet. Polym. Mater.* **2007**, *17*, 483–508. [CrossRef]

26. Pratibha, R.; Kumar, S.; Raina, K.K. Effect of dispersion of gold nanoparticles on the optical and electrical properties of discotic liquid crystal. *Liq. Cryst.* **2014**, *41*, 933–939.

27. García-Cámara, B.; Algorri, J.; Urruchi, V.; Sánchez-Pena, J. Directional scattering of semiconductor nanoparticles embedded in a liquid crystal. *Materials* **2014**, *7*, 2784–2794. [CrossRef] [PubMed]

28. Lee, G.S.; Lee, J.H.; Kim, J.C.; Yoon, T.-H.; Kim, J.-H.; Yu, J.-H.; Choi, H.-Y. Nanoparticle doped in-cell retarder for low operating voltage in transflective liquid crystal displays. *Jpn. J. Appl. Phys.* **2009**, *48*, 042405. [CrossRef]

29. García-García, A.; Vergaz, R.; Algorri, J.F.; Quintana, X.; Otón, J.M. Electrical response of liquid crystal cells doped with multi-walled carbon nanotubes. *Beilstein J. Nanotechnol.* **2015**, *6*, 396–403. [CrossRef] [PubMed]

30. García-García, A.; Vergaz, R.; Algorri, J.F.; Zito, G.; Cacace, T.; Marino, A.; Otón, J.M.; Geday, M.A. Reorientation of single-wall carbon nanotubes in negative anisotropy liquid crystals by an electric field. *Beilstein J. Nanotechnol.* **2016**, *7*, 825–833. [CrossRef] [PubMed]

31. García-García, A.; Vergaz, R.; Algorri, J.F.; Geday, M.A.; Otón, J.M. The peculiar electrical response of liquid crystal-carbon nanotube systems as seen by impedance spectroscopy. *J. Phys. D Appl. Phys.* **2015**, *48*, 375302. [CrossRef]

32. García-Cámara, B.; Gómez-Medina, R.; Sáenz, J.J.; Sepúlveda, B. Sensing with magnetic dipolar resonances in semiconductor nanospheres. *Opt. Express* **2013**, *21*, 23007–23020. [CrossRef]

33. Algorri, J.F.; García-Cámara, B.; García-García, A.; Urruchi, V.; Sánchez-Pena, J.M. Theoretical modeling of a Localized Surface Plasmon Resonance (LSPR) based fiber optic temperature sensor. In Proceedings of the OFS2014 23rd International Conference on Optical Fiber Sensors, Santander, Spain, 2–6 June 2014; p. 915738.

34. Zografopoulos, D.C.; Beccherelli, R. Plasmonic variable optical attenuator based on liquid-crystal tunable stripe waveguides. *Plasmonics* **2013**, *8*, 599–604. [CrossRef]

35. Zografopoulos, D.C.; Beccherelli, R. Long-range plasmonic directional coupler switches controlled by nematic liquid crystals. *Opt. Express* **2013**, *21*, 8240. [CrossRef]

36. Zografopoulos, D.C.; Beccherelli, R. Design of a vertically coupled liquid-crystal long-range plasmonic optical switch. *Appl. Phys. Lett.* **2013**, *102*, 101103. [CrossRef]

37. He, S.; Liang, W.; Cheng, K.-L.; Fang, J.; Wu, S.-T. Bile acid-surfactant interactions at the liquid crystal/aqueous interface. *Soft Matter* **2014**, *10*, 4609–4614. [CrossRef]

38. Otón, E.; Otón, J.M.; Caño-García, M.; Escolano, J.M.; Quintana, X.; Geday, M.A. Rapid detection of pathogens using lyotropic liquid crystals. *Opt. Express* **2019**, *27*, 10098. [CrossRef]

39. Carrasco-Vela, C.; Quintana, X.; Otón, E.; Geday, M.A.; Otón, J.M. Security devices based on liquid crystals doped with a colour dye. *Opto-Electron. Rev.* **2011**, *19*, 496–500. [CrossRef]

40. Marcos, C.; Sánchez-Pena, J.M.; Torres, J.C.; Santos, J.I. Temperature-frequency converter using a liquid crystal cell as a sensing element. *Sensors* **2012**, *12*, 3204–3214. [CrossRef]

41. Algorri, J.F.; Zografopoulos, D.C.; Tapetado, A.; Poudereux, D.; Sánchez-Pena, J.M. Infiltrated photonic crystal fibers for sensing applications. *Sensors* **2018**, *18*, 4263. [CrossRef]

42. Li, J.; Gauza, S.; Wu, S.-T.; Alkeskjold, T.T.; Lægsgaard, J.; Bjarklev, A. High dn o/dT liquid crystals and their applications in a thermally tunable liquid crystal photonic crystal fiber. *Mol. Cryst. Liq. Cryst.* **2006**, *453*, 355–370. [CrossRef]

43. Jang, E. Electrooptic temperature sensor based on a Fabry-Pe/spl acute/rot resonator with a liquid crystal film. *IEEE Photonics Technol. Lett.* **2006**, *18*, 905–907.

44. Wang, Y.; Yang, M.; Wang, D.N.; Liao, C.R. Selectively infiltrated photonic crystal fiber with ultrahigh temperature sensitivity. *IEEE Photonics Technol. Lett.* **2011**, *23*, 1520–1522. [CrossRef]

45. Davies, D.J.D.; Vaccaro, A.R.; Morris, S.M.; Herzer, N.; Schenning, A.P.H.J.; Bastiaansen, C.W.M. A printable optical time-temperature integrator based on shape memory in a chiral nematic polymer network. *Adv. Funct. Mater.* **2013**, *23*, 2723–2727. [CrossRef]

46. Algorri, J.F.; Urruchi, V.; Bennis, N.; Sánchez-Pena, J.M. Liquid crystal temperature sensor based on a micrometric structure and a metallic nanometric layer. *IEEE Electron Device Lett.* **2014**, *35*, 666–668.

47. Algorri, J.F.; Urruchi, V.; Bennis, N.; Sánchez-Pena, J.M. A novel high-sensitivity, low-power, liquid crystal temperature sensor. *Sensors* **2014**, *14*, 6571–6883. [CrossRef]

48. Algorri, J.F.; Garcia-Camara, B.; Urruchi, V.; Sanchez-Pena, J.M. High-sensitivity fabry-pérot temperature sensor based on liquid crystal doped with nanoparticles. *IEEE Photonics Technol. Lett.* **2015**, *27*, 292–295. [CrossRef]

49. Algorri, J.F.; Garcia-Camara, B.; Garcia-Garcia, A.; Urruchi, V.; Sánchez-Pena, J.M. Fiber optic temperature sensor based on amplitude modulation of metallic and semiconductor nanoparticles in a liquid crystal mixture. *J. Light. Technol.* **2015**, *33*, 2451–2455. [CrossRef]

50. Otón, J.M.; Otón, E.; Quintana, X.; Geday, M.A. Liquid-crystal phase-only devices. *J. Mol. Liq.* **2018**, *267*, 469–483. [CrossRef]

51. Bao-Guang, J.; Zhao-Liang, C.; Quan-Quan, M. Simulated human eye retina adaptive optics imaging system based on a liquid crystal on silicon device. *Chin. Phys. B* **2008**, *17*, 4529–4532. [CrossRef]

52. Valley, P.; Reza Dodge, M.; Schwiegerling, J.; Peyman, G.; Peyghambarian, N. Nonmechanical bifocal zoom telescope. *Opt. Lett.* **2010**, *35*, 2582–2584. [CrossRef]

53. Otón, E.; Carrasco, A.; Vergaz, R.; Otón, J.M.; Sánchez-Pena, J.M.; Quintana, X.; Geday, M. 2D tunable beam steering-lens device based on high birefringence liquid crystals. In Proceedings of the 2011 International Conference on Space Optical Systems and Applications (ICSOS), Santa Monica, CA, USA, 11–13 May 2011.

54. Kotova, S.P.; Patlan, V.V.; Samagin, S.A.; Zayakin, O.A. Wavefront formation using modal liquid-crystal correctors. *Phys. Wave Phenom.* **2010**, *18*, 96–104. [CrossRef]

55. Davies, R.; Kasper, M. Adaptive Optics for Astronomy. *Annu. Rev. Astron. Astrophys.* **2012**, *50*. [CrossRef]

56. Kao, Y.-Y.; Huang, Y.-P.; Yang, K.-X.; Chao, P.C.-P.; Tsai, C.-C.; Mo, C.-N. An auto-stereoscopic 3D display using tunable liquid crystal lens array that mimics effects of grin lenticular lens array. *SID Symp. Dig. Tech. Pap.* **2009**, *40*, 111–115. [CrossRef]

57. Algorri, J.F.; Urruchi del Pozo, V.; Sanchez-Pena, J.M.; Oton, J.M. An autostereoscopic device for mobile applications based on a liquid crystal microlens array and an OLED display. *J. Disp. Technol.* **2014**, *10*, 713–720. [CrossRef]

58. Algorri, J.F.; Urruchi, V.; Bennis, N.; Sánchez-Pena, J.M. Cylindrical liquid crystal microlens array with rotary axis and tunable capability. *IEEE Electron Device Lett.* **2015**, *36*, 582–584. [CrossRef]

59. Algorri, J.F.; Urruchi, V.; Bennis, N.; Morawiak, P.; Sanchez-Pena, J.M.; Oton, J.M. Integral imaging capture system with tunable field of view based on liquid crystal microlenses. *IEEE Photonics Technol. Lett.* **2016**, *28*, 1854–1857. [CrossRef]

60. Algorri, J.F.; Urruchi, V.; Bennis, N.; Sánchez-Pena, J.M.; Otón, J.M. Tunable liquid crystal cylindrical micro-optical array for aberration compensation. *Opt. Express* **2015**, *23*, 13899–13915. [CrossRef] [PubMed]

61. Algorri, J.F.; Urruchi, V.; Bennis, N.; Sánchez-Pena, J.M. Modal liquid crystal microaxicon array. *Opt. Lett.* **2014**, *39*, 3476–3479. [CrossRef]

62. Algorri, J.F.; Love, G.D.; Urruchi, V. Modal liquid crystal array of optical elements. *Opt. Express* **2013**, *21*, 24809–24818. [CrossRef] [PubMed]

63. Algorri, J.F.; Bennis, N.; Urruchi, V.; Morawiak, P.; Sánchez-Pena, J.M.; Jaroszewicz, L.R. Tunable liquid crystal multifocal microlens array. *Sci. Rep.* **2017**, *7*, 17318. [CrossRef]

64. Algorri, J.F.; Urruchi, V.; Bennis, N.; Morawiak, P.; Sánchez-Pena, J.M.; Otón, J.M. Liquid crystal spherical microlens array with high fill factor and optical power. *Opt. Express* **2017**, *25*, 605. [CrossRef]

65. Algorri, J.F.; Bennis, N.; Herman, J.; Kula, P.; Urruchi, V.; Sánchez-Pena, J.M. Low aberration and fast switching microlenses based on a novel liquid crystal mixture. *Opt. Express* **2017**, *25*, 14795. [CrossRef] [PubMed]

66. Albero, J.; Garcia-Martinez, P.; Bennis, N.; Oton, E.; Cerrolaza, B.; Moreno, I.; Davis, J.A. Liquid crystal devices for the reconfigurable generation of optical vortices. *J. Light. Technol.* **2012**, *30*, 3055–3060. [CrossRef]

67. Algorri, J.F.; Urruchi, V.; Garcia-Camara, B.; Sanchez-Pena, J.M. Generation of optical vortices by an ideal liquid crystal spiral phase plate. *IEEE Electron Device Lett.* **2014**, *35*, 856–858. [CrossRef]

68. Caño-García, M.; Quintana, X.; Otón, J.M.; Geday, M.A. Dynamic multilevel spiral phase plate generator. *Sci. Rep.* **2018**, *8*, 15804. [CrossRef]

69. Kotova, S.P.; Mayorova, A.M.; Samagin, S.A. Formation of ring-shaped light fields with orbital angular momentum using a modal type liquid crystal spatial modulator. *J. Opt.* **2018**, *20*, 055604. [CrossRef]

70. Algorri, J.; Urruchi, V.; García-Cámara, B.; Sánchez-Pena, J. Liquid crystal lensacons, logarithmic and linear axicons. *Materials* **2014**, *7*, 2593–2604. [CrossRef] [PubMed]

71. Wu, S.-T.; Ren, H. *Introduction to Adaptive Lenses*; Wiley: Hoboken, NJ, USA, 2012; ISBN 9781118270080.

72. Alvarez, L.W. Two-element variable-power spherical lens. US Patent No. 3,305,294, 21 February 1967.

73. Alvarez, L.W. Development of variable- focus lenses and a new refractor. *J. Am. Optom. Assoc.* **1978**, *49*, 24–29. [PubMed]

74. Lohmann, A.W. A new class of varifocal lenses. *Appl. Opt.* **1970**, *9*, 1669. [CrossRef] [PubMed]

75. Barton, I.M.; Dixit, S.N.; Summers, L.J.; Thompson, C.A.; Avicola, K.; Wilhelmsen, J. Diffractive alvarez lens. *Opt. Lett.* **2000**, *25*, 1. [CrossRef]

76. Zhou, G.; Yu, H.; Chau, F.S. Microelectromechanically-driven miniature adaptive Alvarez lens. *Opt. Express* **2013**, *21*, 1226. [CrossRef] [PubMed]

77. Bawart, M.; Jesacher, A.; Zelger, P.; Bernet, S.; Ritsch-Marte, M. Modified Alvarez lens for high-speed focusing. *Opt. Express* **2017**, *25*, 29847. [CrossRef]

78. Schneider, F.; Draheim, J.; Kamberger, R.; Waibel, P.; Wallrabe, U. Optical characterization of adaptive fluidic silicone-membrane lenses. *Opt. Express* **2009**, *17*, 11813. [CrossRef]

79. Zhang, D.-Y.; Lien, V.; Berdichevsky, Y.; Choi, J.; Lo, Y.-H. Fluidic adaptive lens with high focal length tunability. *Appl. Phys. Lett.* **2003**, *82*, 3171–3172. [CrossRef]

80. Ren, H.; Fox, D.; Anderson, P.A.; Wu, B.; Wu, S.-T. Tunable-focus liquid lens controlled using a servo motor. *Opt. Express* **2006**, *14*, 8031. [CrossRef]

81. Sugiura, N.; Morita, S. Variable-focus liquid-filled optical lens. *Appl. Opt.* **1993**, *32*, 4181. [CrossRef]

82. Ren, H.; Wu, S.-T. Variable-focus liquid lens by changing aperture. *Appl. Phys. Lett.* **2005**, *86*, 211107. [CrossRef]

83. Hasan, N.; Banerjee, A.; Kim, H.; Mastrangelo, C.H. Tunable-focus lens for adaptive eyeglasses. *Opt. Express* **2017**, *25*, 1221. [CrossRef] [PubMed]

84. Xiao, W.; Hardt, S. An adaptive liquid microlens driven by a ferrofluidic transducer. *J. Micromechanics Microengineering* **2010**, *20*, 055032. [CrossRef]

85. Cheng, H.-C.; Xu, S.; Liu, Y.; Levi, S.; Wu, S.-T. Adaptive mechanical-wetting lens actuated by ferrofluids. *Opt. Commun.* **2011**, *284*, 2118–2121. [CrossRef]

86. Malouin Jr, B.A.; Vogel, M.J.; Olles, J.D.; Cheng, L.; Hirsa, A.H. Electromagnetic liquid pistons for capillarity-based pumping. *Lab Chip* **2011**, *11*, 393–397. [CrossRef]

87. Ren, H.; Wu, S.-T.; Ren, H.; Wu, S.-T. Adaptive lenses based on soft electroactive materials. *Appl. Sci.* **2018**, *8*, 1085. [CrossRef]

88. Tabiryan, N.; Serak, S.; Dai, X.-M.; Bunning, T. Polymer film with optically controlled form and actuation. *Opt. Express* **2005**, *13*, 7442. [CrossRef] [PubMed]

89. Yu, Y.; Nakano, M.; Ikeda, T. Directed bending of a polymer film by light. *Nature* **2003**, *425*, 145. [CrossRef]

90. Xu, S.; Ren, H.; Lin, Y.-J.; Moharam, M.G.J.; Wu, S.-T.; Tabiryan, N. Adaptive liquid lens actuated by photo-polymer. *Opt. Express* **2009**, *17*, 17590. [CrossRef] [PubMed]

91. Dong, L.; Agarwal, A.K.; Beebe, D.J.; Jiang, H. Adaptive liquid microlenses activated by stimuli-responsive hydrogels. *Nature* **2006**, *442*, 551–554. [CrossRef]

92. Ehrick, J.D.; Stokes, S.; Bachas-Daunert, S.; Moschou, E.A.; Deo, S.K.; Bachas, L.G.; Daunert, S. Chemically tunable lensing of stimuli-responsive hydrogel microdomes. *Adv. Mater.* **2007**, *19*, 4024–4027. [CrossRef]

93. Das, T.K.; Prusty, S. Review on conducting polymers and their applications. *Polym. Plast. Technol. Eng.* **2012**, *51*, 1487–1500. [CrossRef]

94. Nemat-Nasser, S.; Wu, Y. Comparative experimental study of ionic polymer–metal composites with different backbone ionomers and in various cation forms. *J. Appl. Phys.* **2003**, *93*, 5255–5267. [CrossRef]

95. Baughman, R.H.; Cui, C.; Zakhidov, A.A.; Iqbal, Z.; Barisci, J.N.; Spinks, G.M.; Wallace, G.G.; Mazzoldi, A.; De Rossi, D.; Rinzler, A.G.; et al. Carbon nanotube actuators. *Science* **1999**, *284*, 1340–1344. [CrossRef]

96. Shian, S.; Diebold, R.M.; Clarke, D.R. Tunable lenses using transparent dielectric elastomer actuators. *Opt. Express* **2013**, *21*, 8669. [CrossRef]

97. Carpi, F.; Frediani, G.; Turco, S.; De Rossi, D. Bioinspired tunable lens with muscle-like electroactive elastomers. *Adv. Funct. Mater.* **2011**, *21*, 4152–4158. [CrossRef]

98. Ali, M.; Ueki, T.; Tsurumi, D.; Hirai, T. Influence of plasticizer content on the transition of electromechanical behavior of PVC gel actuator. *Langmuir* **2011**, *27*, 7902–7908. [CrossRef]

99. Xia, H.; Takasaki, M.; Hirai, T. Actuation mechanism of plasticized PVC by electric field. *Sens. Actuators A Phys.* **2010**, *157*, 307–312. [CrossRef]

100. Cheng, C.-C.; Andrew Yeh, J. Dielectrically actuated liquid lens. *Opt. Express* **2007**, *15*, 7140. [CrossRef] [PubMed]

101. Kim, Y.; Francl, J.; Taheri, B.; West, J.L. A method for the formation of polymer walls in liquid crystal/polymer mixtures. *Appl. Phys. Lett.* **1998**, *72*, 2253–2255. [CrossRef]

102. Berreman, D.W. Variable-focus LC-lens system. US Patent No. 4,190,330, 26 February 1980.

103. Sato, S. Liquid-crystal lens-cell with variable focal length. *Jpn. J. Appl. Phys.* **1979**, *18*, 1679–1684. [CrossRef]

104. Kowel, S.T.; Cleverly, D.S. A liquid crystal adaptive lens. In Proceedings of the NASA Conference on Optical Information Processing for Aerospace Applications, Hampton, VI, USA, 1981.

105. Cleverly, D. Creation of a Lens by Field Controlled Variation of the Index of Refraction in a Liquid Crystal. Ph.D. Thesis, Syracuse University, Syracuse, NY, USA, 1982.

106. Riza, N.A.; Dejule, M.C. Three-terminal adaptive nematic liquid-crystal lens device. *Opt. Lett.* **1994**, *19*, 1013–1015. [CrossRef]

107. Nose, T.; Sato, S. A liquid crystal microlens obtained with a non-uniform electric field. *Liq. Cryst.* **1989**, *5*, 1425–1433. [CrossRef]

108. Nose, T.; Sato, S. Optical properties of a liquid crystal microlens. In Proceedings of the International Conference on Optoelectronic Science and Engineering, Beijing, China, 22–25 August 1990; Volume 1230.

109. Williams, G.; Powell, N.; Purvis, A.; Clark, M.G. Electrically controllable liquid crystal fresnel lens. In Proceedings of the Opto-electronics Symposium SPIE 1168, San Diego, CA, USA, 7–11 August 1989.

110. Abramochkin, E.G.; Vasiliev, A.A.; Vashurin, P.V.; Zhmurova, L.I.; Ignatov, V.A.; Naumov, A.F. Controlled liquid crystal lens. *Prepr. P. N. Lebedev Phys. Inst.* **1988**, *194*, 18.

111. Naumov, A.F.; Loktev, M.Y.; Guralnik, I.R.; Vdovin, G.V. Liquid-crystal adaptive lenses with modal control. *Opt. Lett.* **1998**, *23*, 992–994. [CrossRef]

112. Love, G.D.; Naumov, A.F. Modal liquid crystal lenses. *Liq. Cryst. Today* **2000**, *10*, 1–4. [CrossRef]

113. Vdovin, G.V.; Guralnik, I.R.; Zayakin, O.A.; Klimov, N.A.; Kotova, S.P.; Loktev, M.Y.; Naumov, A.F.; Patlan, V.V.; Samagin, S.A. Modal liquid crystal wavefront corrector. *Bull. Russ. Acad. Sci.* **2008**, *72*, 71–77.

114. Kotova, S.P.; Patlan, V.V.; Samagin, S.A. Tunable liquid-crystal focusing device. 1. Theory. *Quantum Electron.* **2011**, *41*, 58–64. [CrossRef]

115. Kotova, S.P.; Patlan, V.V.; Samagin, S.A. Tunable liquid-crystal focusing device. 2. Experiment. *Quantum Electron.* **2011**, *41*, 65–70. [CrossRef]

116. Hands, P.J.W.; Tatarkova, S.A.; Kirby, A.K.; Love, G.D. Modal liquid crystal devices in optical tweezing: 3D control and oscillating potential wells. *Opt. Express* **2006**, *14*, 4525. [CrossRef]

117. Korobtsov, A.; Kotova, S.; Losevsky, N.; Mayorova, A.; Samagin, S. Compact optical tweezer with the capability of dynamic control. *J. Biomed. Photonics Eng.* **2015**, *1*, 154–163. [CrossRef]

118. Flexible Optical B., V. LC Lenses. Available online: www.okotech.com/lc. (accessed on 24 May 2019).

119. Xu, S.; Li, Y.; Liu, Y.; Sun, J.; Ren, H.; Wu, S.-T.; Xu, S.; Li, Y.; Liu, Y.; Sun, J.; et al. Fast-response liquid crystal microlens. *Micromachines* **2014**, *5*, 300–324. [CrossRef]

120. Algorri, J.; Urruchi, V.; García-Cámara, B.; Sánchez-Pena, J.; Algorri, J.F.; Urruchi, V.; García-Cámara, B.; Sánchez-Pena, J.M. Liquid crystal microlenses for autostereoscopic displays. *Materials* **2016**, *9*, 36. [CrossRef] [PubMed]

121. Kim, S.-U.; Na, J.-H.; Kim, C.; Lee, S.-D. Design and fabrication of liquid crystal-based lenses. *Liq. Cryst.* **2017**, 1–12. [CrossRef]

122. Bailey, J.; Morgan, P.; Gleeson, H.; Jones, J.; Bailey, J.; Morgan, P.B.; Gleeson, H.F.; Jones, J.C. Switchable liquid crystal contact lenses for the correction of presbyopia. *Crystals* **2018**, *8*, 29. [CrossRef]

123. Lin, Y.-H.; Wang, Y.-J.; Reshetnyak, V. Liquid crystal lenses with tunable focal length. *Liq. Cryst. Rev.* **2017**, *5*, 111–143. [CrossRef]

124. Ye, M.; Wang, B.; Uchida, M.; Yanase, S.; Takahashi, S.; Sato, S. Focus tuning by liquid crystal lens in imaging system. *Appl. Opt.* **2012**, *51*, 7630–7635. [CrossRef]

125. Lin, Y.-H.; Chen, M.-S. A pico projection system with electrically tunable optical zoom ratio adopting two liquid crystal lenses. *J. Disp. Technol.* **2012**, *8*, 401–404. [CrossRef]

126. Lin, H.-C.; Collings, N.; Chen, M.-S.; Lin, Y.-H. A holographic projection system with an electrically tuning and continuously adjustable optical zoom. *Opt. Express* **2012**, *20*, 27222–27229. [CrossRef]

127. Tsou, Y.-S.; Chang, K.-H.; Lin, Y.-H. A droplet manipulation on a liquid crystal and polymer composite film as a concentrator and a sun tracker for a concentrating photovoltaic system. *J. Appl. Phys.* **2013**, *113*, 244504. [CrossRef]

128. Chen, M.; Chen, C.-H.; Lai, Y.; Lu, Y.; Lin, Y.-H. An electrically tunable polarizer for a fiber system based on a polarization-dependent beam size derived from a liquid crystal lens. *IEEE Photonics J.* **2014**, *6*, 1–8. [CrossRef]

129. Kawamura, M.; Ye, M.; Sato, S. Optical trapping and manipulation system using liquid-crystal lens with focusing and deflection properties. *Jpn. J. Appl. Phys.* **2005**, *44*, 6098–6100. [CrossRef]

130. Hassanfiroozi, A.; Huang, Y.-P.; Javidi, B.; Shieh, H.-P.D. Hexagonal liquid crystal lens array for 3D endoscopy. *Opt. Express* **2015**, *23*, 971–981. [CrossRef]

131. Wang, Y.-J.; Chen, P.-J.; Liang, X.; Lin, Y.-H. Augmented reality with image registration, vision correction and sunlight readability via liquid crystal devices. *Sci. Rep.* **2017**, *7*, 433. [CrossRef]

132. Li, G.; Mathine, D.L.; Valley, P.; Ayräs, P.; Haddock, J.N.; Giridhar, M.S.; Williby, G.; Schwiegerling, J.; Meredith, G.R.; Kippelen, B.; et al. Switchable electro-optic diffractive lens with high efficiency for ophthalmic applications. *Proc. Natl. Acad. Sci. USA* **2006**, *103*, 6100–6104. [CrossRef]

133. Peyghambarian, N.; Li, G. Liquid Crystal lenses for correction of presbyopia. In *Adaptive Optics for Industry and Medicine*; Christopher Dainty, Ed.; National University of Ireland: Dublin, Ireland, 2007; Volume 1, pp. 3–8.

134. Milton, H.E.; Morgan, P.B.; Clamp, J.H.; Gleeson, H.F. Electronic liquid crystal contact lenses for the correction of presbyopia. *Opt. Express* **2014**, *22*, 8035. [CrossRef] [PubMed]

135. Kaur, S.; Kim, Y.-J.; Milton, H.; Mistry, D.; Syed, I.M.; Bailey, J.; Novoselov, K.S.; Jones, J.C.; Morgan, P.B.; Clamp, J.; et al. Graphene electrodes for adaptive liquid crystal contact lenses. *Opt. Express* **2016**, *24*, 8782. [CrossRef]

136. Optics, P. Electronic Eyewear. Available online: www.pixeloptics.com/pages/electronic_eyewear. (accessed on 25 June 2015).

137. Lin, Y.-H.; Chen, H.-S. Electrically tunable-focusing and polarizer-free liquid crystal lenses for ophthalmic applications. *Opt. Express* **2013**, *21*, 9428–9436. [CrossRef]

138. Syed, I.M.; Kaur, S.; Milton, H.E.; Mistry, D.; Bailey, J.; Morgan, P.B.; Jones, J.C.; Gleeson, H.F. Novel switching mode in a vertically aligned liquid crystal contact lens. *Opt. Express* **2015**, *23*, 9911. [CrossRef] [PubMed]

139. Bailey, J.; Kaur, S.; Morgan, P.B.; Gleeson, H.F.; Clamp, J.H.; Jones, J.C. Design considerations for liquid crystal contact lenses. *J. Phys. D Appl. Phys.* **2017**, *50*, 485401. [CrossRef]

140. Vanhaverbeke, C.; Verplancke, R.; De Smet, J.; Cuypers, D.; De Smet, H. Microfabrication of a spherically curved liquid crystal display enabling the integration in a smart contact lens. *Displays* **2017**, *49*, 16–25. [CrossRef]

141. Dou, H.; Chu, F.; Guo, Y.-Q.; Tian, L.-L.; Wang, Q.-H.; Sun, Y.-B. Large aperture liquid crystal lens array using a composited alignment layer. *Opt. Express* **2018**, *26*, 9254. [CrossRef]

142. Hsu, C.-J.; Jhang, J.-J.; Huang, C.-Y. Large aperture liquid crystal lens with an imbedded floating ring electrode. *Opt. Express* **2016**, *24*, 16722. [CrossRef]

143. Hsu, C.-J.; Jhang, J.-J.; Jhang, J.-C.; Huang, C.-Y. Influence of floating-ring-electrode on large-aperture liquid crystal lens. *Liq. Cryst.* **2018**, *45*, 40–48. [CrossRef]

144. Li, L.; Bryant, D.; Bos, P.J. Liquid crystal lens with concentric electrodes and inter-electrode resistors. *Liq. Cryst. Rev.* **2014**, *2*, 130–154. [CrossRef]

145. Beeckman, J.; Yang, T.-H.; Nys, I.; George, J.P.; Lin, T.-H.; Neyts, K. Multi-electrode tunable liquid crystal lenses with one lithography step. *Opt. Lett.* **2018**, *43*, 271. [CrossRef]

146. Naumov, A.F.; Love, G.D.; Loktev, M.Y.; Vladimirov, F.L. Control optimization of spherical modal liquid crystal lenses. *Opt. Express* **1999**, *4*, 344–352. [CrossRef]

147. Loktev, M.Y.; Belopukhov, V.N.; Vladimirov, F.L.; Vdovin, G.V.; Love, G.D.; Naumov, A.F. Wave front control systems based on modal liquid crystal lenses. *Rev. Sci. Instrum.* **2000**, *71*, 3290–3297. [CrossRef]

148. Kirby, A.K.; Hands, P.J.; Love, G.D. Liquid crystal multi-mode lenses and axicons based on electronic phase shift control. *Opt. Express* **2007**, *15*, 13496–13501. [CrossRef]

149. Fraval, N.; de Bougrenet, J.L. Low aberrations symmetrical adaptive modal liquid crystal lens with short focal lengths. *Appl. Opt.* **2010**, *49*, 2778–2783. [CrossRef] [PubMed]

150. Hassanfiroozi, A.; Huang, Y.-P.; Javidi, B.; Shieh, H.-P.D. Dual layer electrode liquid crystal lens for 2D/3D tunable endoscopy imaging system. *Opt. Express* **2016**, *24*, 8527. [CrossRef] [PubMed]

151. Ye, M.; Chen, X.; Li, Q.; Zeng, J.; Yu, S. Depth from defocus measurement method based on liquid crystal lens. *Opt. Express* **2018**, *26*, 28413. [CrossRef] [PubMed]

152. Li, H.; Peng, J.; Pan, F.; Wu, Y.; Zhang, Y.; Xie, X. Focal stack camera in all-in-focus imaging via an electrically tunable liquid crystal lens doped with multi-walled carbon nanotubes. *Opt. Express* **2018**, *26*, 12441. [CrossRef]

153. Galstian, T.; Asatryan, K.; Presniakov, V.; Zohrabyan, A.; Tork, A.; Bagramyan, A.; Careau, S.; Thiboutot, M.; Cotovanu, M. High optical quality electrically variable liquid crystal lens using an additional floating electrode. *Opt. Lett.* **2016**, *41*, 3265. [CrossRef] [PubMed]

154. Galstian, T.; Sova, O.; Asatryan, K.; Presniakov, V.; Zohrabyan, A.; Evensen, M. Optical camera with liquid crystal autofocus lens. *Opt. Express* **2017**, *25*, 29945. [CrossRef] [PubMed]

155. Sova, O.; Galstian, T. Liquid crystal lens with optimized wavefront across the entire clear aperture. *Opt. Commun.* **2019**, *433*, 290–296. [CrossRef]

156. Chigrinov, V.; Chigrinov, G.V. Photoaligning and photopatterning—A new challenge in liquid crystal photonics. *Crystals* **2013**, *3*, 149–162. [CrossRef]

157. Sergan, T.A.; Sergan, V.; Herrera, R.; Lu, L.; Bos, P.J.; Sergan, E.V. *In situ* control of surface molecular order in liquid crystals using a localised polymer network and its application to electro-optical devices. *Liq. Cryst.* **2013**, *40*, 72–82. [CrossRef]

158. Sergan, V.V.; Sergan, T.A.; Bos, P.J. Control of the molecular pretilt angle in liquid crystal devices by using a low-density localized polymer network. *Chem. Phys. Lett.* **2010**, *486*, 123–125. [CrossRef]

159. Bezruchenko, V.S.; Muravsky, A.A.; Murauski, A.A.; Stankevich, A.I.; Mahilny, U.V. Tunable liquid crystal lens based on pretilt angle gradient alignment. *Mol. Cryst. Liq. Cryst.* **2016**, *626*, 222–228. [CrossRef]

160. Bezruchenko, V.S.; Mahilny, U.V.; Stankevich, A.I.; Muravsky, A.A.; Murauski, A.A. New photo-crosslinkable benzaldehyde polymers for creating liquid crystal lenses. *J. Appl. Spectrosc.* **2018**, *85*, 704–709. [CrossRef]

161. Kim, J.; Li, Y.; Miskiewicz, M.N.; Oh, C.; Kudenov, M.W.; Escuti, M.J. Fabrication of ideal geometric-phase holograms with arbitrary wavefronts. *Optica* **2015**, *2*, 958. [CrossRef]

162. Lee, Y.-H.; Tan, G.; Zhan, T.; Weng, Y.; Liu, G.; Gou, F.; Peng, F.; Tabiryan, N.V.; Gauza, S.; Wu, S.-T. Recent progress in Pancharatnam–Berry phase optical elements and the applications for virtual/augmented realities. *Opt. Data Process. Storage* **2017**, *3*, 79–88. [CrossRef]

163. Ma, Y.; Tam, A.M.W.; Gan, X.T.; Shi, L.Y.; Srivastava, A.K.; Chigrinov, V.G.; Kwok, H.S.; Zhao, J.L. Fast switching ferroelectric liquid crystal Pancharatnam-Berry lens. *Opt. Express* **2019**, *27*, 10079. [CrossRef]

164. Zhan, T.; Lee, Y.-H.; Tan, G.; Xiong, J.; Yin, K.; Gou, F.; Zou, J.; Zhang, N.; Zhao, D.; Yang, J.; et al. Pancharatnam–Berry optical elements for head-up and near-eye displays. *J. Opt. Soc. Am. B* **2019**, *36*, D52. [CrossRef]

165. Moon, S.; Lee, C.-K.; Nam, S.-W.; Jang, C.; Lee, G.-Y.; Seo, W.; Sung, G.; Lee, H.-S.; Lee, B. Augmented reality near-eye display using Pancharatnam-Berry phase lenses. *Sci. Rep.* **2019**, *9*, 6616. [CrossRef] [PubMed]

166. Zhou, Y.; Yin, Y.; Yuan, Y.; Lin, T.; Huang, H.; Yao, L.; Wang, X.; Tam, A.M.W.; Fan, F.; Wen, S. Liquid crystal Pancharatnam-Berry phase lens with spatially separated focuses. *Liq. Cryst.* **2018**, 1–6. [CrossRef]

167. Jamali, A.; Bryant, D.; Zhang, Y.; Grunnet-Jepsen, A.; Bhowmik, A.; Bos, P.J. Design of a large aperture tunable refractive Fresnel liquid crystal lens. *Appl. Opt.* **2018**, *57*, B10. [CrossRef]

168. Srivastava, A.K.; Wang, X.Q.; Gong, S.Q.; Shen, D.; Lu, Y.Q.; Chigrinov, V.G.; Kwok, H.S. Micro-patterned photo-aligned ferroelectric liquid crystal Fresnel zone lens. *Opt. Lett.* **2015**, *40*, 1643. [CrossRef] [PubMed]

169. Kumar, M.B.; Kang, D.; Jung, J.; Park, H.; Hahn, J.; Choi, M.; Bae, J.-H.; Kim, H.; Park, J. Ultrathin, polarization-independent, and focus-tunable liquid crystal diffractive lens for augmented reality. *arXiv* **2019**, arXiv:1902.10889.

170. Wei, X.; Zheng, J.; Wang, Y.; Gao, Z.; Sun, L.; Lu, Y.; Zhuang, S. Multi-imaging characteristics of electrically controlled on-axis holographic polymer-dispersed liquid-crystal Fresnel lens. *Opt. Eng.* **2015**, *54*, 037110. [CrossRef]

171. Lin, S.-H.; Li, C.-Y.; Kuo, C.-T.; Yeh, H.-C. Fresnel lenses in 90° twisted-nematic liquid crystals with optical and electrical controllability. *IEEE Photonics Technol. Lett.* **2016**, *28*, 1462–1464. [CrossRef]

172. Lin, S.-H.; Huang, B.-Y.; Li, C.-Y.; Yu, K.-Y.; Chen, J.-L.; Kuo, C.-T. Electrically and optically tunable Fresnel lens in a liquid crystal cell with a rewritable photoconductive layer. *Opt. Mater. Express* **2016**, *6*, 2229. [CrossRef]

173. Noda, K.; Momosaki, R.; Kawai, K.; Sakamoto, M.; Sasaki, T.; Kawatsuki, N.; Goto, K.; Ono, H. Trifocal lens system with liquid crystal Fresnel lens. *Jpn. J. Appl. Phys.* **2018**, *57*, 102502. [CrossRef]

174. Wang, X.-Q.; Yang, W.-Q.; Liu, Z.; Duan, W.; Hu, W.; Zheng, Z.-G.; Shen, D.; Chigrinov, V.G.; Kwok, H.-S. Switchable Fresnel lens based on hybrid photo-aligned dual frequency nematic liquid crystal. *Opt. Mater. Express* **2017**, *7*, 8. [CrossRef]

175. Avci, N.; Lee, Y.-H.; Hwang, S.-J. Switchable polarisation-independent blue phase liquid crystal Fresnel lens based on phase-separated composite films. *Liq. Cryst.* **2017**, *44*, 1078–1085. [CrossRef]

176. Lin, H.-Y.; Avci, N.; Hwang, S.-J. High-diffraction-efficiency Fresnel lens based on annealed blue-phase liquid crystal–polymer composite. *Liq. Cryst.* **2019**, 1–8. [CrossRef]

177. Rong, N.; Li, Y.; Li, X.; Zhou, P.; Liu, S.; Lu, J.; Su, Y. Polymer-stabilized blue-phase liquid crystal fresnel lens cured with patterned light using a spatial light modulator. *J. Disp. Technol.* **2016**, *12*, 1008–1012. [CrossRef]

178. Dou, H.; Chu, F.; Wang, L.; Tian, L.-L.; Li, R.; Hou, W.-Y.; Wang, Q.-H. A polarisation-free blue phase liquid crystal lens with enhanced tunable focal length range. *Liq. Cryst.* **2018**, 1–7. [CrossRef]

179. Chen, H.-S.; Wang, Y.-J.; Chang, C.-M.; Lin, Y.-H. A polarizer-free liquid crystal lens exploiting an embedded-multilayered structure. *IEEE Photonics Technol. Lett.* **2015**, *27*, 899–902. [CrossRef]

180. Zhan, T.; Xiong, J.; Lee, Y.-H.; Wu, S.-T. Polarization-independent Pancharatnam-Berry phase lens system. *Opt. Express* **2018**, *26*, 35026. [CrossRef] [PubMed]

MDPI

St. Alban-Anlage 66

4052 Basel

Switzerland

Tel. +41 61 683 77 34

Fax +41 61 302 89 18

www.mdpi.com

Crystals Editorial Office

E-mail: crystals@mdpi.com

www.mdpi.com/journal/crystals